Engineering Materials and Processes

Series Editor

Professor Brian Derby, Professor of Materials Science
Manchester Materials Science Centre, Grosvenor Street, Manchester, M1 7HS, UK

William W. Sampson

Modelling Stochastic Fibrous Materials with *Mathematica*®

Springer

William W. Sampson, PhD
School of Materials
University of Manchester
Sackville Street
Manchester
M60 1QD
UK

ISBN 978-1-84996-811-9   e-ISBN 978-1-84800-991-2

DOI 10.1007/978-1-84800-991-2

Engineering Materials and Processes ISSN 1619-0181

A catalogue record for this book is available from the British Library

Cover design: eStudio Calamar S.L., Girona, Spain

Printed on acid-free paper

9 8 7 6 5 4 3 2 1

springer.com

Preface

This is a book with three functions. Primarily, it serves as a treatise on the structure of stochastic fibrous materials with an emphasis on understanding how the properties of fibres influence those of the material. For some of us, the structure of fibrous materials is a topic of interest in its own right, and we shall see that there are many features of the structure that can be characterised by rather elegant mathematical treatments. The interest of most researchers however is the manner in which the structure of fibrous networks influences their performance in some application, such as the ability of a non-woven textile to capture particles in a filtration process or the propensity of cells to proliferate on an electrospun fibrous scaffold in tissue engineering. The second function of this book is therefore to provide a family of mathematical techniques for modelling, allowing us to make statements about how different variables influence the structure and properties of materials. The final intended function of this book is to demonstrate how the software *Mathematica*[1] can be used to support the modelling work, making the techniques accessible to non-mathematicians.

We proceed by assuming no prior knowledge of any of the main aspects of the approach. Specifically, the text is designed to be accessible to any scientist or engineer whatever their experience of stochastic fibrous materials, mathematical modelling or *Mathematica*. We begin with an introduction to each of these three topics, starting with defining clearly the criteria that classify stochastic fibrous materials. We proceed to consider the reasons for using models to guide our understanding and specifically why *Mathematica* has been chosen as a computational aid to the process.

Importantly, this book is intended not just to be read, but to be *used*. It has been written in a style intended to allow the reader to extract all the important relationships from the models presented without using the *Mathematica*

---

[1] A trial version of *Mathematica* 6.0 can be downloaded from http://www.wolfram.com/books/resources, by entering the licence number L3250-9882. *Mathematica* notebook files containing the code presented in each chapter can be downloaded from http://www.springer.com/978-1-84800-990-5.

examples provided. However, for a comprehensive understanding of the science underlying the models, readers will benefit from running the code and editing it to probe further the dependencies we identify and discuss. The *Mathematica* code embedded in the text includes numbers added by *Mathematica* on evaluation. Where a line of code begins with 'In[1]:=' this indicates that a new session on the *Mathematica* kernel has been started and previously defined variables, *etc.* no longer exist in the system memory.

When preparing the manuscript, some consideration was given as to the appropriate amount of *Mathematica* code to include. In common with many scientists, the author's first approach to any theoretical analysis involves the traditional tools of paper and pencil, with the use of *Mathematica* being introduced after the initial formulation of the problem of interest. Typically however, once progress has been made on a problem, the early manipulations, *etc.* are subsequently entered into *Mathematica* so that the full treatment is contained in a single file. As a rule, we seek to replicate this approach by providing many of the preliminary relationships and straightforward manipulations as ordinary typeset equations before turning to *Mathematica* for the more demanding aspects of the analysis. Occasionally, where the outputs of relatively simple manipulations are required for subsequent treatments, *Mathematica* is invoked at an earlier stage.

Consideration was given also as to whether to include *Mathematica* code for the generation of plots, or to provide these only in the form of figures with supplementary detail, such as arrows, *etc.* One of the great advantages of working with *Mathematica* is that its advanced graphics capabilities allow rapid generation of plots and surfaces representing functions; surfaces can be rotated using a mouse or other input device and dynamic plots with interactivity are readily generated. When developing theory, the ability to visualise functions guides the process and can provide valuable reassurance that functions behave in a way that is representative of the physical system of interest. Accordingly, the *Mathematica* code used to generate graphics is provided in the majority of instances and graphics are not associated with figure numbers, but instead are shown as the output of a *Mathematica* evaluation. Where graphics are associated with a figure number, the content is either a drawing to guide our analysis or a collection of results from several *Mathematica* evaluations; in the latter case, plots or surfaces have been typically generated using *Mathematica* and exported to graphics software for the addition of supplementary detail and annotation.

The manuscript has benefited greatly from the comments of Kit Dodson, Nicola Dooley, Ramin Farnood and Steve I'Anson, who provided helpful feedback on an early draft. I would like to thank each of them for being so generous of their time and for being so diligent in their attention to detail. Kit Dodson deserves particular thanks and recognition for introducing me to statistical geometry, stochastic modelling and *Mathematica* when I spent time with his research group at the University of Toronto in the early 1990s. Many of the outcomes from our fifteen years of fruitful and enjoyable research collaboration are included in this monograph.

Thanks are also due to Maryka Baraka of Wolfram Research for support and advice, to Steve Eichhorn for permission to reproduce the micrograph of an electrospun nanofibrous network in Figure 1.1 and to Steve Keller for permission to reproduce Figures 5.7 and 5.8. I would like to thank Taylor and Francis Ltd. for permission to reproduce Figure 3.7, Journal of Pulp and Paper Science for permission to reproduce Figure 6.1 and Wiley-VCH Verlag GmbH & Co. KGaA for permission to reproduce Figure 7.1.

Manchester   *Bill Sampson*
February, 2008

Contents

1

Introduction

Probably the oldest and most familiar stochastic fibrous material to most of us is that on which this text is printed. Tradition has it that paper was invented in China at the start of the second century CE though there is some evidence for its existence as early as the second century BCE. Regardless of the precise date that paper saw its first use, we may be confident that it is not as old as the material from which it takes its name, papyrus. These two materials, both of which have been so important in recording the development of society and ideas, provide a good pair of examples with which we can guide our classification of materials as stochastic. Papyrus is made from the pithy inner part of the stem of the sedge *Cyperus papyrus*; during the manufacture of papyrus, this is cut into long strips that are laid side by side on a flat surface with their edges overlapping. A second layer with the strips oriented perpendicularly to those in the first layer is placed over the first. On drying, these strips bond to each other to yield the sheet-like material we call papyrus. So, during the manufacture of papyrus, each strip is carefully placed in relation to those strips which have already been laid down and, given knowledge of the location and orientation of one strip, we can be quite confident of the location and orientation of others. This regularity in the structure classifies it as *deterministic* and other materials with structures encompassed by this classification include woven textiles, honeycomb structures, *etc*.

Consider now the structure of paper. This material is made by the filtration of an aqueous suspension of fibres over a woven mesh so that the fibres are retained on the mesh. The resultant fibrous filter cake is pressed to remove water retained between the fibres bringing them into intimate contact such that, after drying, usually under heat, the fibres bond to each other at these regions of contact. Evidently, the filtration process does not yield the control of the location of fibres in the sheet that can be exerted over the location of strips in papyrus. Indeed, whereas for papyrus we may be confident of the location and orientation of any strip given this information about another, for paper we cannot predict the location and orientation of any given fibre given the same information about another. We shall see that although it is possible

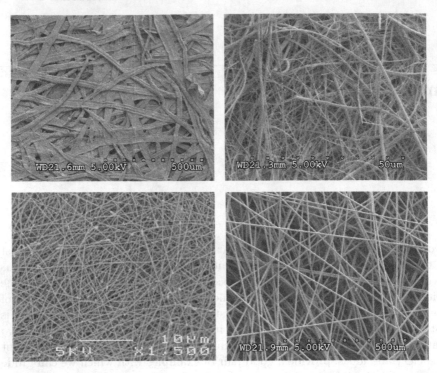

Figure 1.1. Micrographs of four planar stochastic fibrous materials. Clockwise from top left: paper formed from softwood fibres, glass fibre filter, non-woven carbon fibre mat, electrospun nylon nanofibrous network (Courtesy S.J. Eichhorn and D.J. Scurr. Reproduced with permission)

to make statements of the *probability* of a given structural feature occurring, the material is neither uniform nor regularly non-uniform, but it is variable in a way that can be characterised by statistics. Accordingly, we classify paper and similar materials as *stochastic*.

Stochastic processes arise in a number of interesting contexts such as the motion of gases, the evolution of bubbles of different sizes in bread and the clustering of traffic on the motorway. Of interest to us is the structure of stochastic fibrous networks; some micrographs of planar stochastic fibrous materials are shown in Figure 1.1. The image on the top left shows the surface of a sheet of paper formed from fibres from a softwood tree; adjacent to this is a higher magnification image of a glass fibre filter paper of the type used in laboratories. On first inspection, we observe that this network consists of fibres with at least two classes of width and that these exhibit more curvature than those in the sheet of paper. Nonetheless, there are evident similarities between these two structures. Both are porous, and the pores formed by the intersections of fibres have a distribution of sizes. Also, despite the curvature of the fibres, the distances between adjacent intersections on any given fibre are sufficiently close that the fibrous ligaments between these can be considered, to a reasonable approximation, as being straight so the inter-fibre voids are rather polygonal.

Two networks formed from very straight fibres are shown on the bottom of Figure 1.1. The image on the right shows a network of carbon fibres used in electromagnetic shielding applications and in the manufacture of gas diffusion layers for use in fuel cells. In common with paper and the glass fibre filter that we have just considered, this material is formed by a process where fibres are deposited from an aqueous suspension. The micrograph reveals another important structural property of the planar stochastic fibrous materials that we have considered so far – the fibres lie very much in the plane of the network and do not exhibit any significant degree of entanglement. We shall consider this further in the sequel, and will utilise this property of materials in developing models characterising their structures. The network on the bottom left of Figure 1.1 shares these characteristics though it is formed by a very different process: electrospinning. Here the fibres from which the network is formed are manufactured in the same process as the network itself and as the filament leaves the spinneret it is subjected to electrostatic forces, elongating it and yielding a layered stochastic fibrous network. There has been considerable interest in such materials in recent years, driven in part by the potential of electrospinning processes to yield fibres of width a few nanometres and fibres that are themselves porous, providing opportunities to tailor structures for application as biomaterials and for nanocomposite reinforcements, see *e.g.* [91].

In Chapters 3 to 6 we will derive models for materials of the type we have discussed so far, where fibre axes can be considered to lie in the plane of the material. These layered structures represent the most common type of fibre network encountered in a range of technical and engineering applications. Stochastic fibrous materials do exist however where fibre axes are oriented in three dimensions. These include needled and hydro-entangled non-woven textiles such as felts and insulating materials and the fibrous architectures within short-fibre reinforced composite materials. We derive models for the structure of this class of materials in Chapter 7.

1.1 Random, Near-Random and Stochastic

So far, we have used the fact that the structure of the fibre networks we are considering varies in a non-deterministic way to classify them as being stochastic. The term stochastic is often used interchangeably with the term *random*. Here we will instead consider random structures to be a special class of stochastic fibrous materials. We will classify a random process as one where the events are independent of each other and equally likely; three criteria

were identified in one of the seminal works on modelling planar random fibre
networks, that of Kallmes and Corte [74]. These are:

- the fibres are deposited independently of one another;
- the fibres have an equal probability of landing at all points in the
  network;
- the fibres have an equal probability of making all possible angles
  with any arbitrarily chosen, fixed axis.

So, from the first two of these criteria, we consider the random events to be
the incidence of fibre centres within an area that represents some or all of our
network. Now, fibre centres exist at points in space, but the third criterion
given by Kallmes and Corte arises because fibres are extended objects, *i.e.*
they have appreciable aspect ratio. In their simplest form, we may consider
fibres as rectangles with their major axes being straight lines. The third criterion tells us that these major axes have equal probability of lying within any
interval of angles, so it effectively restates the first two criteria for lines rather
than points—the angle made by the major axis of a fibre is independent of
those of other fibres and all angles are equally likely. When fibres are not
straight but exhibit some curvature along their length, then the same definition holds, but for the third criterion we consider that the tangents to the
major axis have an equal probability of making all possible angles with any
arbitrarily chosen, fixed axis. Note that for materials such as those shown in
Figure 1.1 we consider orientation of fibre axes to be effectively in the plane,
whereas for three-dimensional networks, orientation is defined by solid angles
in the range 0 to 2π steradians. Tomographic images of networks of this type
are shown in Figure 7.1.

Graphical representations of random point and line processes generated
using *Mathematica* are shown in Figure 1.2. The graphic on the left of Figure 1.2 shows 1,000 points which we will consider to represent the centres
of fibres. These points are generated independently of each other with equal
probability that they lie within any region of the unit square. On first inspection, we note a very important property of random processes—they are
not regularly spaced but are clustered, *i.e.* they exhibit clumping. The extent
of this clustering is a characteristic of the process that depends upon its intensity only, so for our process of fibre centres in the unit square it depends
upon the number of fibre centres per unit area. The graphic on the right of
Figure 1.2 uses the coordinates of the points shown in the graphic on the left
as the centres of lines with length 0.1 and with uniformly random orientation
as stipulated in the third criterion of Kallmes and Corte. The interaction of
lines with each other provides connectivity to the network and increases our
perception of its non-uniformity through, for example, the intensity of the
line process within different regions of the unit square or the different sizes of
polygons bounded by these lines in the same regions.

In the following chapters, we shall make use of mathematics to provide
a quantitative framework for our definition of random fibre networks but for

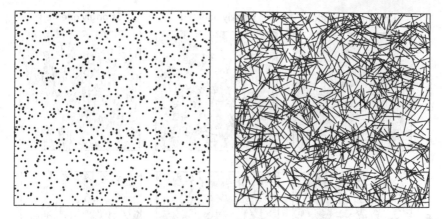

Figure 1.2. Random point and fibre processes in two dimensions. Left: 1,000 random points in a unit square; right: the same points extended to be random lines with the points at their centre and with length 0.1

now we state simply that a random fibre process is a specific class of process that exactly satisfies the criteria of Kallmes and Corte mathematically. When networks are made in the laboratory or in an industrial manufacturing context, the resultant structures typically display some differences from those generated by model random processes, *i.e.* system influences combine to yield networks with structures that are manifestly not deterministic, so we still require the use of statistics to describe them, yet also they do not meet the precise mathematical criteria that permit them to be classified as random. We will characterise such influences as yielding 'departures from randomness', and these fall into three principal categories:

- Preferential orientation of fibres to a given direction,
- Fibre clumping,
- Fibre dispersion.

Graphical representations of these departures from randomness are given in Figure 1.3 for networks of 1,000 lines of uniform length 0.1 with centres occurring within a unit square. The structure on the left of the second row represents a model random network.

The preferential orientation of fibres to a given direction often arises as a consequence of the inherent directionality of manufacturing processes. Most fibrous materials are manufactured in continuous processes that result in the reeling of a web on a roll; a consequence of this is that fibres are typically delivered in suspension of air or water to the forming section of the machine with a highly directional flow and are thus oriented preferentially in the direction of manufacture. We will consider models for the influence of fibre orientation in Chapter 6.

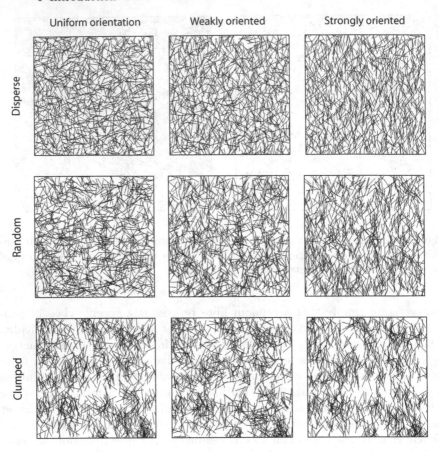

Figure 1.3. Departures from randomness. Industrially formed networks typically exhibit different degrees of fibre clustering and fibre orientation than the model random fibre network shown on the left of the second row

Our definition of a random fibre network requires that fibres are deposited independently of each other. In Chapter 7 we will determine the limiting volumetric concentration of fibres required for them to be independent of each other in three-dimensions, but for now we note that at the concentrations used in almost all industrial web-forming processes and many laboratory forming processes, fibres are in contact with several other fibres such that they form clumps or *flocs*. Accordingly, fibres are not deposited independently of each other, but their centres are more likely to be found close to another fibre centre than in the random case, resulting in a more clumpy structure. Fibre clumping is more common in stochastic fibrous materials than fibre dispersion, which yields a structure with less clustering than a model random network, but which is still stochastically variable.

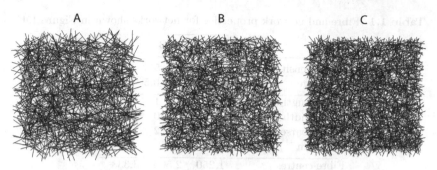

Figure 1.4. Influence of fibre properties on structures of random fibre networks. Graphic rendered to allow fibre length to extend beyond the square region containing fibre centres

Whereas a preferential orientation of fibres in a network is potentially rather easy to observe, the extent of clumping or dispersion in the network is not so readily detected. In part this is because the extent of clumping in a network depends upon the geometry and morphology of its constituent fibres. We illustrate this in Figure 1.4, which shows three simulated random networks of straight uniform fibres; each network has mass per unit area, 5 g m^{-2} and consists of fibre centres occurring as a uniform random process in a square of side 1 cm. The mass per unit area of fibrous materials is often termed the *grammage* of the network, or its *areal density*. Three properties of fibres were required to generate the structures shown in Figure 1.4: the length and width of fibres and their mass per unit length; this property of fibres is termed *coarseness* or *linear density* and is often reported for textiles and textile filaments in units of denier, or grams per 9,000 metres. The fibre and network properties for the networks shown in Figure 1.4 are given in Table 1.1, which introduces also the network variable *coverage*, which we define as the number of fibres covering a point in the plane of support of the network; the average coverage of the network is given by its mass per unit area divided by that of the constituent fibres. Since the coarseness of a fibre is defined as its mass per unit length, it follows that its mass per unit area is given by its coarseness divided by its width, so the coverage is given by

$$\text{Coverage} = \frac{\text{Network areal density} \times \text{Fibre width}}{\text{Fibre coarseness}}.$$

We shall encounter coverage as a stochastic variable rather frequently in the following chapters; for now we note that it is a dimensionless number and that by holding this constant for the networks rendered in Figure 1.4 we ensure that the networks retain a further common feature: the expected fraction of the network that is covered by fibres is the same for the three cases presented.

On first inspection of the Figure 1.4 it is immediately apparent that the networks are rather different. We can make the qualitative observation that

Table 1.1. Fibre and network properties for networks shown in Figure 1.4

	A	B	C
Areal density $(\mathrm{g\,m^{-2}})$	5	5	5
Coverage	0.5	0.5	0.5
Fibre length (mm)	2	1	1
Fibre width (μm)	20	20	15
Fibre coarseness $(10^{-4}\mathrm{g\,m^{-1}})$	2	2	1.5
Fibre centres per unit area $(\mathrm{cm^{-2}})$	1,250	2,500	3,333
Total fibre length per unit area $(\mathrm{cm^{-1}})$	250	250	333

the uniformity of the networks improves as we move from left to right. Referring to Table 1.1 we note that the fibres in Network A are twice as long as those in Network B. Thus, to achieve the same areal density and coverage, Network B consists of twice as many fibres as Network A. Accordingly, the total fibre length per unit area in Networks A and B is the same, yet their structures differ through the number of fibres per unit area and the extent of interaction of a given fibre with others in the network; we expect the latter to be a function of fibre length. Similarly, Network C exhibits greater uniformity than Network B because it has greater fibre length per unit area as a consequence of the linear density of the constituent fibres being less.

It is clear that the structural characteristics of a random fibre network depend upon the properties of the constituent fibres. We are already in a position then to make some recommendations as to how we might influence the uniformity of a fibrous material by choice of fibres with given dimensions. For example, we may state that network uniformity can be improved by increasing the total fibre length per unit area in the network and by reducing fibre length. In due course, we shall see that many properties of random fibre networks can be expressed explicitly in terms of the variables that we considered in interpreting Figure 1.4, *i.e.* the length, width and coarseness of fibres, and the mass per unit area and coverage of the network. So, instead of applying general rules we seek to be in a position to state precisely the influence of changing one of these variables on some quantitative descriptor of uniformity or, for example, on the extent of inter-fibre contact or mean void size of the network. It is important to bear in mind however, that to a lesser or greater extent, real networks exhibit the departures from randomness discussed earlier and illustrated in Figure 1.3. Accordingly, we classify randomness as a special case, and consider random fibre networks as a reference structure against which measurements made on real structures can be compared. We will define precisely functions that characterise random processes and that permit us to calculate the properties of random fibre networks. We will therefore characterise the full family of fibrous networks governed by spatial distributions of fibres with a distribution of orientations as 'stochastic'; the term 'random' is used for model structures the conform to the criteria of Corte and Kallmes, as introduced on page 4. Where departures from randomness are weak, we will classify our structures as 'near-random'. No strict demarcations exist between these terms, and we shall see that whilst some properties of a given stochastic material may differ significantly from the random case, in the same sample others may be remarkably close.

1.2 Reasons for Theoretical Analysis

Before looking at the reasons for using *Mathematica* to assist the modelling process, it is worth considering why modelling *per se* is a useful tool when applying the scientific method. A good mathematical model provides expressions that describe the behaviour of a system faithfully in terms of the variables that influence it. A good example of such a model is the equation for the period, T (s) of a simple pendulum, which most of us have encountered:

$$T = 2\pi\sqrt{\frac{l}{g}} \,, \tag{1.1}$$

where l (m) is the length of the pendulum from its pivot point to the centre of mass of the bob and g (m s^{-2}) is the acceleration due to gravity. The derivation of Equation 1.1 depends on some assumptions:

- the bob is a point mass;
- the string or rod on which the bob swings has no mass;
- the motion of the pendulum occurs in a plane;
- the pendulum exists in a vacuum;
- the initial angle of the string or rod to the vertical, θ, is sufficiently small that $\theta \approx \sin(\theta)$.

It turns out that Equation 1.1 describes the motion of a pendulum rather well for initial angular displacements of about 20° or less, so for such systems we can be confident that the only sensible way to influence the period of the pendulum is to change the length of the string or rod from which it is suspended and that the period is proportional to the square root of this length. This is a simplified representation of a real system. Of course, we can derive expressions for systems with larger angular displacements, or elliptical orbits, *etc.* but the differential equations required to account for these do not lend themselves to closed form solutions and require numerical integration. The important issue here is the *usefulness* of the expressions available. Because we are aware of the assumptions made in deriving Equation 1.1, we are aware also of the range of conditions where we may expect it to provide an accurate prediction of the period of a pendulum. If we wish to study a system where the assumptions made in deriving Equation 1.1 do not apply, then we must use a more complicated model, and probably must work a little harder to make good predictions of its behaviour.

Of course, Equation 1.1 is a simple expression describing a simple system. Indeed, by conducting a series of experiments measuring the periods of pendulums with different weight bobs, string lengths and initial angular displacements, it is likely that after some informed data processing we might come up with a similar equation by heuristic methods. Such experiments are useful to theoreticians in their own right – they provide data against which we can compare our theories in order that we can verify them, or, if agreement is not as good as we would like, the differences guide the development of more faithful models – they are, however, time consuming and most systems are significantly more complex than the simple pendulum.

Now, there is uncertainty associated with any experimental measurement and we might characterise this by reporting the spread of experimental data about the mean using, for example, confidence intervals. For a deterministic process, such as the swinging of a simple pendulum, the spread of data captures variability due to experimental error, instrument accuracy, *etc.* The materials of interest to us are stochastic and, by definition, there is additional variability in measurements of their properties that is not due to experimental error or uncertainty, but which is a characteristic of the material. Accordingly, we will seek to model the likelihood that the networks exhibit certain properties and, for the properties that exhibit dependence on many variables, theory allows us to identify which of these have the more significant influence on the property of interest. In stochastic fibre networks we do not know the location or orientation of any individual component such as a fibre, pore or inter-fibre contact, relative to others in the structure. We must therefore use statistics to describe their combined effect and, provided the number of components is sufficiently large, theory will accurately describe the properties of the network as a whole.

Although the mass of the bob does not appear as a variable in Equation 1.1, this mass is present in its derivation through energy considerations; thus, bearing the assumptions in mind, the experimentalist should be guided by the theory to avoid seeking to influence the period of the pendulum through changing the experimental variables of mass or initial displacement. So, theory can be used to guide practical and experimental work and this is particularly useful for complex stochastic systems. In Chapter 5 we will derive expressions for the mean in-plane pore dimension in random fibre networks and will see that this is influenced by fibre width and network porosity but not by fibre length. In Chapter 6 we will see that clumping, as a departure from randomness, *is* influenced by fibre length but has only a weak influence on pore size. Thus, to a researcher investigating the use of porous fibrous scaffolds for tissue engineering, we would recommend that they influence the pore size of their networks in the laboratory by changing fibre width and porosity, and the experiment can be guided by the theory.

As well as guiding practical work and our thinking, theory can be used to provide insights which are rather difficult to obtain even in the controlled environment of the laboratory. Consider the linear density of fibres, which we encountered on page 7 and defined as their mass per unit length. A fibre of circular cross section with width ω and density, ρ, has linear density given by

$$\delta = \frac{\pi\,\omega^2}{4}\,\rho\ .\tag{1.2}$$

So we see that linear density is proportional to the square of fibre width such that if we alter fibre width in an experimental setting, then we expect to influence linear density also. Thus we would be unable to determine without further experimentation whether an observed behaviour is due to the change in width or the change in linear density. Of course, we might carry out experiments using hollow fibres, where changing the thickness of the fibre wall allows us to change fibre width without changing linear density. Obtaining such fibres, with sufficiently well characterised geometries may prove rather difficult, but the experiment could be carried out. Using a theoretical approach however, we can include fibre width and linear density as *independent* variables in a model, so that their influence on the property of interest can be decoupled. Accordingly, models conserve and focus experimental effort.

The final reason for theoretical analysis that we consider here is that equations enable us to model systems outside the realms permitted by experiments. Thus, whilst it may be difficult to obtain an experimental measure of, for example, the tortuosity of paths from one side of a fibre network to the another, an expression for this property is rather easy to derive for an isotropically porous material (*cf.* Section 5.5). In short, our models should represent applied mathematics in a form that is both useful and useable. Throughout the following chapters, we seek to simplify our analyses to provide accessible and tractable expressions aiding their application.

1.3 Modelling with *Mathematica*

We have noted that mathematical modelling is applied mathematics that should be useful and useable. The models that we derive should therefore be accessible to scientists and engineers who are likely to have studied disciplines other than mathematics in arriving at their given expertise. Although scientists and engineers with an interest in modelling can be expected to have a good background in calculus, statistics, *etc.*, it is not uncommon for us to find that we have sufficient mathematical knowledge to formulate a problem, but an insufficient knowledge of the full range of techniques available to advance a solution; *Mathematica* helps us to overcome this problem. To an extent it serves as a helpful mathematician that looks after the integrals, *etc.* that we may not know how to handle. In fulfilling this role, it turns out that

Mathematica serves as rather a good teacher also. Consider, for example, the integral

$$\int \frac{e^x}{x}\, \mathrm{d}x\ .$$

We input this into *Mathematica* as

In[1]:= **Integrate[e^x / x, x]**

where the second argument of the command **Integrate** specifies that we want to integrate our function with respect to x. We press $\boxed{\text{Shift}}$ + $\boxed{\text{Enter}}$ together to evaluate our input and obtain the output:

Out[1]= ExpIntegralEi [x]

By selecting this output and pressing the F1 key, we access the *Mathematica* help files for **ExpIntegralEi** we identify the function as the exponential integral function. This is an example of a 'special function' that few of us other than mathematicians will have encountered in formal studies. We will encounter it later, but what is important for now is the fact that *Mathematica* will handle most functions that we present to it and that it may provide us with answers with which we are unfamiliar. Importantly, through the help files and exploration of these functions we rapidly become familiar with them. So we may do, for example,

In[2]:= **Plot[ExpIntegralEi[x], {x, -2, 2}]**

Out[2]=

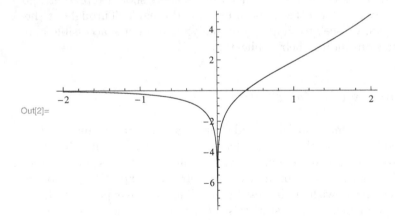

The syntax is rather intuitive; the first argument to the command **Plot** is the function we wish to plot and the second argument specifies that we want to plot the function for values of x between -2 and 2.

In fact, we could ask *Mathematica* to carry out the same integral using several different notations. For example,

In[3]:= **Integrate[Exp[x] / x, x]**

In[4]:= $\int \dfrac{e^x}{x}\, dx$

In[5]:= **Integrate[E^x / x, x]**

are all equivalent. This is important. Different users prefer different styles of input; accordingly, the code presented in the chapters that follow represents just one way of approaching the problems we consider. *Mathematica* has a large array of palettes allowing symbolic input of expressions such that they closely resemble what we write with a pencil and paper. Typically, the *Mathematica* code presented here will have been composed with commands given as words rather than as symbols, so we shall use **Integrate** instead of $\int$, **Sum** instead of $\sum$, *etc.* Regardless of the chosen style of input, the computation will be symbolic, rather than numerical, unless stated otherwise.

In addition to its strong symbolic capabilities, *Mathematica* provides an excellent environment for experimental mathematics. In particular, the command **Manipulate** provides interactivity which can guide our thinking. We illustrate this by considering a simple example where we investigate the range of x for which the approximation $\sin(x) \approx x$ is reasonable. In our example, the first argument of the **Manipulate** command specifies the function of interest; the second argument specifies that we seek to vary parameter x between 0 and π with starting value $x = 1$:

In[6]:= **Manipulate[{x, Sin[x] / x}, {{x, 1.}, 0, π}]**

Out[6]=
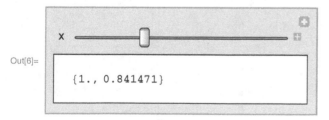
{1., 0.841471}

The output includes a slider that can be used to vary the value of parameter x and yield the bracketed term $\{x, \sin(x)/x\}$ in the output field.

The static format of a book necessarily restricts ready application of dynamic objects such as those generated using **Manipulate** and thus we do not make extensive use of the command in our examples. We will make use of the command **Table** to generate lists and nested lists of data and objects however, and readers may readily add interactivity to their *Mathematica* code by applying the rule-of-thumb that dynamic objects may often be created by replacing the command **Table** with **Manipulate**.

As we begin our treatments in the following chapters, no knowledge of *Mathematica* is assumed and the use of individual commands and the appropriate use of syntax will be introduced. As we proceed however, readers should find the choice of commands more intuitive and less explanation is provided. Throughout, plots have been generated with line styles chosen to allow good monochrome reproduction; when working on a computer, many of the options specified using the argument **PlotStyle** will be redundant, as *Mathematica* automatically renders graphics with readily distinguishable colours for different functions, data-sets, *etc.*

2

Statistical Tools and Terminology

2.1 Introduction

Having classified our materials as being stochastic, we require a family of mathematical tools to represent the distributions of their properties and some suitable numbers to describe these distributions. This chapter provides informally some background to these tools. A real number is called a 'random variable' if its value is governed by a well-defined statistical distribution. We begin by defining some general properties of random variables and many of the distributions that we will encounter in subsequent chapters and that we shall use to derive the properties of stochastic fibrous materials. As well as using standard mathematical notation, the use of *Mathematica* to handle statistical functions and generate random data is introduced.

2.2 Discrete and Continuous Random Variables

We have identified the difference between stochastic and deterministic processes as being essentially one of uncertainty. Often this uncertainty arises because we do no know enough about the factors that contribute to the state of the process or its outcome. Consider for example the rolling of a fair six-sided die. If we knew enough about the position, orientation in three-dimensions, and velocity of the die at some given point in time, as well as the relevant elastic moduli and coefficients of friction of the die and the surface onto which we are rolling, then we might develop appropriate equations of motion and solve these to compute the precise position at rest of the die and hence predict the number that will be rolled. This is a difficult problem to formulate, let alone solve even if all the equations and variables were known; typically we expect that at least the first three will be unknown. Accordingly, we have uncertainty in our system. In fact, even if we create a machine to roll the die identically for several throws, we expect that different outcomes will result because of the sensitivity to even small uncertainties in the variables. We are unable therefore to deterministically predict the outcome of a roll and must always be uncertain of any individual event. Despite this uncertainty, we may be confident that the probability of rolling any number is $\frac{1}{6}$. Thus, whereas we cannot predict the outcome of an individual roll, we know what all the possible outcomes are and the probability of their occurrence. We can state then the random variable x which represents the outcome of the roll of a die can take the values 1, 2, 3, 4, 5 and 6 and each outcome has probability $\frac{1}{6}$. In the sequel, we shall see that this characterises the random variable x as being controlled by the discrete uniform probability distribution, $P(x) = \frac{1}{6}$.

We consider first the application of statistics to the description of systems where the events within that system or the outcomes of it are *discrete*. This means that each possible event or outcome has a definite probability of occurrence. We have just considered one such process, the rolling of an unbiased die. Another example of a discrete stochastic process is the tossing of a coin where the probability of the outcome being either heads or tails is $\frac{1}{2}$. If we assume that the probability of the die coming to rest on one of its edges is infinitesimal, then we may state that the probability of each event is $\frac{1}{6}$. Similarly, we know that it is not possible to throw the die and have the uppermost face show, for example, $4\frac{1}{2}$ spots. So the outcome of rolling the die is a discrete random variable. Examples of discrete random variables that characterise the structure of fibre networks are the number of fibre centres per unit volume or area in the network, or the number of fibres making contact with any given fibre in the structure. As a rule, we can expect to encounter discrete random variables when the feature of interest, experimental conditions permitting, may be counted; the exception to this being where only certain classes of events exist, for example, where a fibre network is formed from a blend of fibres manufactured with precisely known lengths which are known because they have been measured and not because they have been counted.

Consider now the distribution of the weights of eggs produced by free-range hens. The probability that an egg weighs precisely 60 g is very small; as is the probability that it weighs precisely 59.9 g or 60.000001 g. It is much easier, and certainly more meaningful, to state the probability that eggs from these hens weigh between say 55 and 65 g or between 45 and 55 g, *etc.* Clearly, the weights of the eggs differ from the rolling of a die in that we do not have discrete outcomes; the weight of an egg is therefore classified as a *continuous* random variable. Examples of continuous random variables encountered in the description of fibre networks are the area or volume of inter-fibre voids and the lengths of the fibrous ligaments that exist between fibre crossings.

2.2.1 Characterising Statistics

Given sample data from a system, *e.g.* the outcomes, x_i of n rolls of a die or the weights of n eggs, we may use statistics to characterise the population. The most common statistics to characterise the distribution are familiar to most of us through the handling of experimental data. We define them here for completeness.

Mean: The mean value of the sample data is given by the sum of all the data divided by the number of observations. For data $x_1, x_2 \ldots x_n$ we denote the mean $\bar{x}$ and this is given by

$$\bar{x} = \sum_{i=1}^{n} \frac{x_i}{n} \ . \tag{2.1}$$

The mean is often termed the *expectation* or, in every-day language, the *average*.

Mode: The mode is the value within our sample data that occurs with the greatest frequency. For discrete data, this is found by inspection; for continuous data the mode is estimated from a histogram of the data as the mid point of the tallest column.

Median: The median is occasionally used instead of the mean for the characterisation of data that has a histogram that is not symmetric about the mean; such data is described as *skewed*. The median is found by sorting the data by magnitude and selecting the middle observation such that half the observations are numerically greater than the median and half are numerically smaller.

Variance: The variance of our data is the mean square difference from the mean, *i.e.* it is the expected value of $(x_i - \bar{x})^2$. It is denoted $\sigma^2(x)$ and given by

$$\sigma^2(x) = \sum_{i=1}^{n} \frac{(x_i - \bar{x})^2}{n} \ . \tag{2.2}$$

For small samples of data, Equation 2.2 will underestimate the variance because, for a sample of size n, each observation can be independently compared with only $(n-1)$ other observations, biasing the calculation of the variance. Accordingly, the unbiased estimate of the variance is given by

$$\sigma^2(x) = \sum_{i=1}^{n} \frac{(x_i - \bar{x})^2}{n - 1} \ , \tag{2.3}$$

and this is typically applied for samples with n less than about 20.

Standard Deviation: The standard deviation is the square root of the variance and it is denoted $\sigma(x)$. It is often preferred to the variance as it has the same units as the original data.

$$\sigma(x) = \sqrt{\sigma^2(x)} = \sqrt{\sum_{i=1}^{n} \frac{(x_i - \bar{x})^2}{n}} \tag{2.4}$$

The unbiased estimate is given by

$$\sigma(x) = \sqrt{\sigma^2(x)} = \sqrt{\sum_{i=1}^{n} \frac{(x_i - \bar{x})^2}{n-1}} \ . \tag{2.5}$$

Coefficient of Variation: The coefficient of variation is the standard deviation relative to the mean. We denote it $CV(x)$ and it is given by

$$CV(x) = \frac{\sigma(x)}{\bar{x}} \ . \tag{2.6}$$

Note that the coefficient of variation is dimensionless and is often reported as a percentage.

We classify the mean, mode and median as measures of location; they provide a measure of the magnitude of the numbers we can expect to characterise our distribution. The variance, standard deviation and coefficient of variation are classified as measures of spread; they provide a measure of how widely distributed the data are in our sample or population and thus can be used to inform how representative our measures of location are of the distribution as a whole.

Using Characterising Statistics

We illustrate the calculation of these characterising statistics with *Mathematica* by generating a sample of data representing rolls of a pair of unbiased dice using the command **RandomInteger**. This function generates pseudorandom integers with equal probability, so the command **RandomInteger[]** will give an output of either 0 or 1 with the probability of each outcome being $\frac{1}{2}$. To represent the roll of a fair six-sided die we use **RandomInteger[1,6]**.

Consider first the outcomes of rolling a pair of unbiased dice 20 times. The outcomes of the experiment are recorded in the following graphic:

In fact, these dice rolls were simulated in *Mathematica* using **RandomInteger** with the following input:

```
SeedRandom[1]
pairs = RandomInteger[{1, 6}, {20, 2}]
```

which gives the output in list form which corresponds to our graphic:

```
Out[2]= {{5, 3}, {5, 1}, {2, 1}, {1, 3}, {1, 1}, {4, 6}, {3, 1},
        {4, 5}, {5, 2}, {4, 4}, {5, 2}, {5, 3}, {2, 2}, {5, 6},
        {5, 6}, {1, 4}, {4, 1}, {1, 3}, {4, 2}, {2, 4}}
```

Note the use of the command **SeedRandom**. By including this line, *Mathematica* uses the same random seed for each evaluation and we obtain the same value for **pairs** each time we evaluate the code. Each pair of numbers is identified in *Mathematica* by its location in the list, so we can refer to these using the command **Part** or the assignment **[[ ]]**, *e.g.* ,

```
In[3]:= Part[pairs, 4]
        pairs[[8]]
```

```
Out[3]= {1, 3}
```

```
Out[4]= {4, 5}
```

The values obtained by summing the numbers shown on each pair of dice represent the random variable of interest. For the ith pair of random numbers, we obtain their sum using **Total[pairs[[i]]]**, and we use the command **Table** to carry this out for all i:

```
In[5]:= rolls = Table[Total[pairs[[i]]], {i, 1, 20}]
```

```
Out[5]= {8, 6, 3, 4, 2, 10, 4, 9, 7, 8, 7, 8, 4, 11, 11, 5, 5, 4, 6, 6}
```

To compute the mean of our dice rolls we need to apply Equation 2.1 and compute the sum of all observations and divide this by the number of observations. *Mathematica* has a built-in command **Mean** to carry out this calculation:

```
In[6]:= Mean[rolls]
```

$$Out[6]= \frac{32}{5}$$

The result is displayed as an improper fraction, because *Mathematica* has carried out computations on random integers. To convert to the corresponding numerical value, we use **N**:

```
In[7]:= N[%]
```

```
Out[7]= 6.4
```

where the symbol **%** refers to the last output. To compute the variance we require the mean square difference from the mean, as given by Equation 2.3. We might compute this explicitly using,

In[8]:= **Total[(rolls - Mean[rolls])^2]/19**
       N[%]

Out[8]= $\dfrac{644}{95}$

Out[9]= 6.77895

though again, *Mathematica* has the specific command **Variance** to handle this for us:

In[10]:= **Variance[rolls]**

Out[10]= $\dfrac{644}{95}$

Inevitably, the standard deviation is given by,

In[11]:= **StandardDeviation[rolls]**
        N[%]

Out[11]= $2\sqrt{\dfrac{161}{95}}$

Out[12]= 2.60364

and is the square root of the variance:

In[13]:= **TrueQ$\left[$StandardDeviation[rolls] == $\sqrt{\text{Variance[rolls]}}\,\right]$**

Out[13]= True

Importantly in Version 6, *Mathematica* always uses Equations 2.3 and 2.5 to calculate the variance and standard deviation when handling lists. Note that to generate the square root operator in *Mathematica* we use $\boxed{\text{Ctrl}}$+2, though we could obtain the square root of the variance using any of the following:

```
Sqrt[Variance[rolls]]
Variance[rolls] ^ (1 / 2)
Variance[rolls]^{1/2}
Power[Variance[rolls], 1 / 2]
```

where in the third example, the superscript is generated using $\boxed{\text{Ctrl}}$ +6.

For completeness, we calculate the remaining measures of location and spread for our data, as given earlier in this section:

```
In[14]:= Median[rolls]
         Commonest[rolls] (* Commonest = mode *)
         CVrolls = StandardDeviation[rolls] / Mean[rolls]
         N[CVrolls]
```

```
Out[14]= 6
```

```
Out[15]= {4}
```

$$Out[16]= \frac{\sqrt{\frac{805}{19}}}{16}$$

```
Out[17]= 0.406819
```

The use of the command **Median** is an intuitive choice, but we note that the command **Mode** is used in *Mathematica* in conjunction with commands associated with equation solving and other operations; thus we compute the mode using the command **Commonest**. Note that the output of this command is a list enclosed in braces, { }, in our case this list has length 1, though this need not be the case. Note also the use of the comment enclosed between starred brackets, (* *); anything between these characters is not evaluated.

If we change the first line of our code to **SeedRandom[2]** we obtain a different set of observations:

```
In[18]:= SeedRandom[2]
         pairs = RandomInteger[{1, 6}, {20, 2}]
         rolls = Table[Total[pairs[[i]]], {i, 1, 20}];
```

```
Out[19]= {{6, 2}, {3, 3}, {6, 3}, {2, 6}, {6, 1}, {1, 5}, {4, 5},
          {1, 2}, {2, 6}, {2, 6}, {5, 5}, {1, 1}, {5, 5}, {2, 3},
          {4, 4}, {1, 2}, {1, 5}, {3, 3}, {2, 6}, {5, 4}}
```

and the output of **rolls** has been suppressed by ending this line of code with a semi-colon. Calculating the mean, variance and standard deviation as before we have,

```
In[21]:= N[Mean[rolls]]
         N[Variance[rolls]]
         N[StandardDeviation[rolls]]

Out[21]= 6.95

Out[22]= 5.31316

Out[23]= 2.30503
```

On first inspection, it is clear that the calculated mean, variance and standard deviation for our two sets of simulated dice rolls are different. This arises because we have only a limited set of data available to characterise the distribution, *i.e.* we are considering the statistics of two *samples* that we hope are representative of the *population* from which they are drawn. Using different values of **SeedRandom** we have generated independent samples from the population of dice rolls where the probabilities of a given number being shown on the face of each dice are equal. Of course, we might pool the results of our two samples to provide a better estimate of the statistics that characterise the distribution:

```
In[24]:= SeedRandom[1]
         pairs = RandomInteger[{1, 6}, {20, 2}];
         rolls1 = Table[Total[pairs[[i]]], {i, 1, 20}];
         SeedRandom[2]
         pairs = RandomInteger[{1, 6}, {20, 2}];
         rolls2 = Table[Total[pairs[[i]]], {i, 1, 20}];
         pooledrolls = Join[rolls1, rolls2];
```

Note here that the name **pairs** is used twice, so values arising from the first evaluation are overwritten in the *Mathematica* kernel by those from the second evaluation. The command **Join** concatenates the specified lists. The characterising statistics for the pooled data are given in the usual way:

```
In[31]:= Length[pooledrolls]
         N[Mean[pooledrolls]]
         N[Variance[pooledrolls]]
         N[StandardDeviation[pooledrolls]]

Out[31]= 40

Out[32]= 6.675

Out[33]= 5.96859

Out[34]= 2.44307
```

and we observe that our new estimate of the mean is precisely the mean of our two estimates from the independent samples. The estimates of the variance, and hence the standard deviation, lie between those of the two samples, but are not the mean of these estimates as they are calculated on the basis of the new estimate of the mean and a larger sample with $n = 40$.

Mathematica can handle very large lists very comfortably, so we get a much improved estimate of the characterising statistics using larger n:

```
In[35]:= SeedRandom[1]
         n = 1 000 000;
         pairs = RandomInteger[{1, 6}, {n, 2}];
         rolls =
           Table[Total[pairs[[i]]], {i, 1, Length[pairs]}];
         Mean[N[rolls]]
         StandardDeviation[N[rolls]]

Out[39]= 7.00089

Out[40]= 2.41501
```

Note the placing of the command **N** such that the calculations are performed on numerical rather than integer values of the random variable. This speeds up the calculations as illustrated by use of the command **Timing**, which gives the output as a list where the first term is the time taken in seconds for *Mathematica* to perform the calculation:

```
In[41]:= StandardDeviation[rolls] // Timing
         N[StandardDeviation[rolls]] // Timing
         StandardDeviation[N[rolls]] // Timing
```

$$\text{Out[41]= } \left\{7.1, \frac{3\sqrt{\frac{375\,016\,153}{37\,037}}}{125}\right\}$$

Out[42]= {8.142, 2.41501}

Out[43]= {0.09, 2.41501}

So for this example, the calculation of the standard deviation is almost 80 times faster when performed numerically.

Using our list of length 1 million, we can track the dependence of our calculation of the mean and standard deviation on the size of our sample. To do this, we compute the mean and standard deviation for samples of increasing length, n using the command **Take** to extract elements from the list and the command **Table** to do this for different n. In the example that follows we compute the mean for samples of length between 100 and 100,000 in steps of 100. The output of **meanrollsn** is a list of sublists, each of length 2, where the first element is the size of the sample, n and the second element is the mean of that sample. Using **ListPlot** we are able to visualise the quality of our estimate of the mean as we increase the sample size.

```
In[44]:= meanrollsn = Table[{n, Mean[N[Take[rolls, n]]]},
            {n, 100, 100 000, 100}];
         ListPlot[meanrollsn, PlotRange → All,
          AxesLabel → {"n", "Mean"}]
```

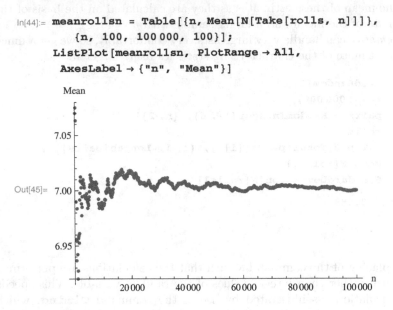

We use similar code to calculate the standard deviation for different n:

```
In[46]:= stdrollsn =
            Table[{n, StandardDeviation[N[Take[rolls, n]]]},
             {n, 100, 100 000, 100}];
         ListPlot[stdrollsn, PlotRange → All,
          AxesLabel → {"n", "Standard deviation"}]
```

From inspection of the graphical outputs generated using **ListPlot** we can be reasonably confident that a sample size of some tens of thousands will provide us with a reasonable estimate of the characterising statistics for our distribution. When dealing with a sample of size 1 million, we might consider that the statistics of our sample approach those of the population. As yet, though, we do not know precisely the characterising statistics for the population from which our samples are drawn. Referring back to our simulation of 1 million rolls, we might reasonably assume that the mean of the population is 7 and the standard deviation is about 2.42. Note that if we used **SeedRandom[2]** to simulate a million rolls of a pair of dice, our estimate of the mean would change in the 4th decimal place, whereas that of the standard deviation would differ in the third. We will now consider how we can use probability theory to obtain robust measures of location and spread for statistical populations.

Theoretical Determination of Characterising Statistics

Numerical approaches of the type used so far are often referred to as Monte Carlo methods and are very useful when theoretical approaches do not lend themselves to closed form solutions. Very often however, statistical theory does allow us to make precise statements about the properties of distributions. We consider first theory describing the problem of rolling a single die and proceed to consider the case of rolling a pair of dice, which we have just considered.

Consider first the rolling of a fair six-sided die. The only possible outcomes are the integers 1 to 6 and each outcome has probability $\frac{1}{6}$. Since the family of possible outcomes is limited to these values, we have a discrete random variable and, since all outcomes have the same probability, our random variable has a *discrete uniform distribution*. For random integers x with $x_{min} \leq x \leq x_{max}$ the probability of a given x_i is given by

$$P(x) = \begin{cases} 0 & \text{if } x < x_{min} \\ \frac{1}{1+x_{max}-x_{min}} & \text{if } x_{min} \leq x \leq x_{max} \\ 0 & \text{otherwise} \end{cases} \qquad (2.7)$$

In *Mathematica* the discrete uniform distribution is input as

In[1]:= **DiscreteUniformDistribution[{xmin, xmax}]**

Out[1]= DiscreteUniformDistribution[{xmin, xmax}]

and the probability function is input using

In[2]:= **PDF[DiscreteUniformDistribution[{xmin, xmax}], x]**

Out[2]= $\dfrac{1}{1 + \text{xmax} - \text{xmin}}$

which corresponds to the second interval of the piecewise function given by Equation 2.7. Note that *Mathematica* is aware of the definition of the distribution for arbitrary x:

```
In[3]:=  PDF[DiscreteUniformDistribution[{1, 6}], x]
         Table[PDF[DiscreteUniformDistribution[{1, 6}], x],
           {x, 0, 8}]
         PDF[DiscreteUniformDistribution[{1, 6}], 2.2]
```

$$\text{Out[3]}= \ \frac{1}{6}$$

$$\text{Out[4]}= \ \left\{0, \ \frac{1}{6}, \ \frac{1}{6}, \ \frac{1}{6}, \ \frac{1}{6}, \ \frac{1}{6}, \ \frac{1}{6}, \ 0, \ 0\right\}$$

$$\text{Out[5]}= \ 0$$

Of course, *Mathematica*'s functions are well defined and have been fully tested. However, when deriving our own probability functions later, we will frequently check that we have accounted for all possible outcomes by ensuring that the probability function sums to 1:

```
In[6]:=  Sum[PDF[DiscreteUniformDistribution[{xmin, xmax}], x],
           {x, xmin, xmax}]
```

$$\text{Out[6]}= \ 1$$

Having reassured ourselves of this, we can compute the mean using

$$\bar{x} = \sum_{x=x_{\min}}^{x_{\max}} x\, P(x) \tag{2.8}$$

Note that whereas when handling data, the mean was calculated as the sum of all observations divided by the number of observations, here we compute the product of the value of the observation x_i and its frequency of occurrence and sum the result for all possible x. We input this as:

```
In[7]:=  xbar = Sum[
           x PDF[DiscreteUniformDistribution[{xmin, xmax}], x],
           {x, xmin, xmax}]
```

$$\text{Out[7]}= \ \frac{\text{xmax} + \text{xmin}}{2}$$

Similarly, to compute the variance we require,

$$\bar{x} = \sum_{x=x_{\min}}^{x_{\max}} (x - \bar{x})^2 P(x) \tag{2.9}$$

which we compute using

In[8]:= $\mathbf{Sum}\Big[(\mathbf{x} - \mathbf{xbar})^2$
$\qquad \mathbf{PDF[DiscreteUniformDistribution[\{xmin, xmax\}], x],}$
$\qquad \mathbf{\{x, xmin, xmax\}}\Big]$

Out[8]= $\dfrac{1}{12} (\text{xmax} - \text{xmin}) (2 + \text{xmax} - \text{xmin})$

For distributions that are predefined in *Mathematica* we can compute these statistics directly, though the output of the command **Variance** requires some manipulation to yield the same form as given by the summing method:

In[9]:= $\mathbf{Mean[DiscreteUniformDistribution[\{xmin, xmax\}]]}$
$\qquad \mathbf{Variance[DiscreteUniformDistribution[\{xmin, xmax\}]]}$
$\qquad \mathbf{Factor[\%]}$

Out[9]= $\dfrac{\text{xmax} + \text{xmin}}{2}$

Out[10]= $\dfrac{1}{12} \left(-1 + (1 + \text{xmax} - \text{xmin})^2 \right)$

Out[11]= $\dfrac{1}{12} (\text{xmax} - \text{xmin}) (2 + \text{xmax} - \text{xmin})$

In the case of our six-sided die, we have $x_{\min} = 1$ and $x_{\max} = 6$ and the mean and variance are given by

In[12]:= $\mathbf{Mean[DiscreteUniformDistribution[\{1, 6\}]]}$
$\qquad \mathbf{Variance[DiscreteUniformDistribution[\{1, 6\}]]}$

Out[12]= $\dfrac{7}{2}$

Out[13]= $\dfrac{35}{12}$

and we observe that the mean, or the expected value, is not a possible outcome. This is an important property of discrete random variables and will shall encounter it in other contexts as we develop theory describing the structure of stochastic fibrous materials.

We return now to the two-dice problem that we considered numerically earlier. The possible outcomes and their probabilities are summarised in Table 2.1. It is immediately clear that the distribution of outcomes is symmetrical about 7, which is the mode of our distribution. We require a probability function that describes our random variable and by inspection we note that

Face value	Permutations	Probability
2	1	$\frac{1}{36}$
3	2	$\frac{2}{36} = \frac{1}{18}$
4	3	$\frac{3}{36} = \frac{1}{12}$
5	4	$\frac{4}{36} = \frac{1}{9}$
6	5	$\frac{5}{36}$
7	6	$\frac{6}{36} = \frac{1}{6}$
8	5	$\frac{5}{36}$
9	4	$\frac{4}{36} = \frac{1}{9}$
10	3	$\frac{3}{36} = \frac{1}{12}$
11	2	$\frac{2}{36} = \frac{1}{18}$
12	1	$\frac{1}{36}$

Table 2.1. Permutations and probabilities for outcomes of rolling a pair of unbiased six-sided dice

the probabilities on the right of our graphic are given by[1]

$$P(x) = \begin{cases} \frac{6-|7-x|}{36} & \text{if } 2 \leq x \leq 12 \\ 0 & \text{otherwise} \end{cases} \qquad (2.10)$$

To input this to *Mathematica* we introduce two new commands. Firstly, instead of assigning a variable name to the function, we use the function **SetDelayed** which we input as **:=** such that the right-hand side of our input is not evaluated until called. We also use the command **Piecewise** to assign probability zero for all x outside the applicable range of our function.

```
In[14]:= P[x_] :=
         Piecewise[{{(6 - Abs[7 - x]) / 36, 2 ≤ x ≤ 12}}, 0]
```

We should check that our probability function yields the required probabilities:

```
In[15]:= Table[P[x], {x, 0, 14}]
```

$$\text{Out[15]= } \left\{0, 0, \frac{1}{36}, \frac{1}{18}, \frac{1}{12}, \frac{1}{9}, \frac{5}{36}, \frac{1}{6}, \frac{5}{36}, \frac{1}{9}, \frac{1}{12}, \frac{1}{18}, \frac{1}{36}, 0, 0\right\}$$

and check also that we have considered all probabilities:

---

[1] In the general case, the random variable $Y = X_1 + X_2$ with $1 \leq X_1, X_2 \leq X_{max}$ where X_1 and X_2 are independent discrete random variables taking integer values, has probability function,

$$P(Y) = \frac{X_{max} - |X_{max} + 1 - Y|}{X_{max}^2} .$$

In[16]:= **Sum[P[x], {x, 2, 12}]**

Out[16]= 1

The mean, variance and standard deviation are given by

In[17]:= **xbar = Sum[x P[x], {x, 2, 12}]**

xvar = Sum[(x - xbar)^2 P[x], {x, 2, 12}]

xstd = $\sqrt{\text{xvar}}$

N[%]

Out[17]= 7

Out[18]= $\dfrac{35}{6}$

Out[19]= $\sqrt{\dfrac{35}{6}}$

Out[20]= 2.41523

By using the probability function for the outcomes of rolling a pair of unbiased-six sided dice, we are able to make precise statements about the characterising statistics of our distribution. The expected outcome, *i.e.* the mean, is 7; since this outcome has the highest probability and the distribution is symmetrical about the mean, the mode and median are 7 also. The standard deviation of the distribution is $\sqrt{35/6}$. We observe that these theoretical measures agree rather closely with those obtained for a million dice rolls.

2.3 Common Probability Functions

In the last section we encountered the discrete uniform distribution and identified the *Mathematica* commands to call this distribution and to generate its probability function, its mean, and its variance. The discrete uniform distribution is one of the simplest distributions we are likely to encounter; we have a finite number of permissible outcomes in an interval, and these have equal probability. Before considering continuous random variables, where the number of outcomes in an interval is infinite, we introduce some more probability functions that characterise the distributions of discrete random variables and which we shall use extensively in modelling the structure of fibrous materials.

2.3.1 Bernoulli Distribution

The Bernoulli distribution is used to characterise random processes where there are only two possible outcomes. The classical example of such a process is the tossing of a coin where the outcomes 'heads' or 'tails' each have probability $\frac{1}{2}$, though other examples include observations by researchers of whether cars travelling in rush-hour are occupied by the driver only or by the driver and passengers or whether a random point within a block of sandstone lies with a void or in the solid phase of the material. In this latter case, the probability that the point lies in a void is its porosity, ϵ and the probability that the point lies within the solid is $(1 - \epsilon)$.

By convention, we denote the outcomes 0 and 1 and often these are taken to classify the outcomes as 'failure' and 'success', respectively. If the probability of success is $0 \leq p \leq 1$, then the probabilities of success and failure are given by

$$P(0) = 1 - p \qquad\qquad (2.11)$$
$$P(1) = p \ , \qquad\qquad (2.12)$$

which can be written as,

$$P(n) = p^n \, (1 - p)^{1-n} \ . \qquad\qquad (2.13)$$

We call the Bernoulli distribution in *Mathematica* using

```
BernoulliDistribution[p]
```

To obtain the probability function we use

In[1]:= `PDF[BernoulliDistribution[p], n]`

Out[1]= $\begin{cases} 1 - p & n == 0 \\ p & n == 1 \end{cases}$

which corresponds to Equation 2.13, but it is expressed in piecewise form. Note that the piecewise function given in the output uses the notation '`==`' for 'equals'; the single equals sign, '`=`', as used so far, allows us to set a value to the variable name preceding it.

The mean and variance of the Bernoulli distribution are given by

In[2]:= `Mean[BernoulliDistribution[p]]`
`Variance[BernoulliDistribution[p]]`

Out[2]= `p`

Out[3]= `(1 - p) p`

The appropriate graphical representation of a discrete probability function is a bar chart, which we generate using the command **BarChart**. This command is not loaded by default when the *Mathematica* kernel is launched, so we must call the required package using the command **Needs**. The following code calls the package **BarCharts** and generates a list of Bernoulli probabilities for the case where the probability of success $p = 0.7$. This list is then plotted as a bar chart with appropriate axis labels. Note that the last line unsets the value of parameter **p**.

```
In[4]:= Needs["BarCharts`"]
        p = .7;
        Pn =
            Table[PDF[BernoulliDistribution[p], n], {n, 0, 1}];
        BarChart[Pn, BarLabels → {0, 1},
          AxesLabel → {"n", "P(n)"}]
        p =.
```

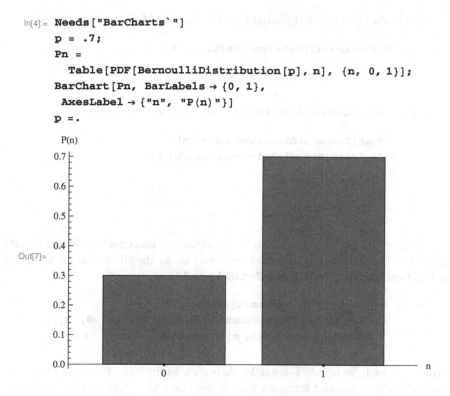

2.3.2 Binomial Distribution

Consider now an extension to the examples we considered when introducing the Bernoulli distribution. If we toss a coin m times, observe m cars to see if they are carrying passengers, or select m points at random from within the volume of a block of sandstone to identify if they are in the solid or void phase, we may be interested in how many of these m *Bernoulli trials* have a particular outcome, *i.e.* how many result in 'success' or 'failure'. The distribution describing this discrete random variable is the binomial distribution. Denoting the number of successes $0 \leq x \leq m$ for Bernoulli trials with probability of success p, it has probability function,

$$P(x) = \binom{m}{x} (1 - p)^{m-x} p^x \tag{2.14}$$

where $\binom{m}{x}$ is the binomial coefficient,

$$\binom{m}{x} = \frac{m!}{x!\,(m-x)!} \tag{2.15}$$

and it is invoked using **Binomial[m, x]** in *Mathematica*.

To obtain the binomial probability function in *Mathematica* we use

In[9]:= **PDF[BinomialDistribution[m, p], x]**

Out[9]= $(1 - p)^{m-x} p^x$ Binomial[m, x]

The mean and variance of the binomial distribution are given by

In[10]:= **Mean[BinomialDistribution[m, p]]**
        Variance[BinomialDistribution[m, p]]

Out[10]= m p

Out[11]= m (1 - p) p

To plot the probability function, we again use **BarChart**. To aid investigation of the influence of parameters p and m on the distribution, we define a function **bar[p_, m_]** using **SetDelayed** (**:=**).

In[12]:= **bar[m_, p_] := BarChart[Table[**
           PDF[BinomialDistribution[m, p], x], {x, 0, m}],
           BarLabels → Range[0, m], AxesLabel → {"x", "P(x)"}]

Note that we have nested several *Mathematica* commands, neatening the code; note also the command **Range** which is used here to generate a list representing the labels on the abscissa. This is required because by default **BarChart** labels the first bar, '1', the second '2', *etc.*, yet for our data, the first bar represents the probability of outcome zero, the second the probability of outcome 1, *etc.*

Now, we could investigate the influence of parameter p, representing the probability of success in a trial, by evaluating, for example,

        bar[10, .2]
        bar[20, .5]

etc. and generate several bar charts for comparison. As we have mentioned earlier however, *Mathematica* allows us to alter variables interactively using the command **Manipulate**. We generate a bar chart of the probability function for the binomial distribution with two associated sliders allowing us to vary p and m using,

In[14]:= **Manipulate[bar[m, p], {{p, .5}, 0, 1}, {m, 10, 100, 5}]**

Out[14]=

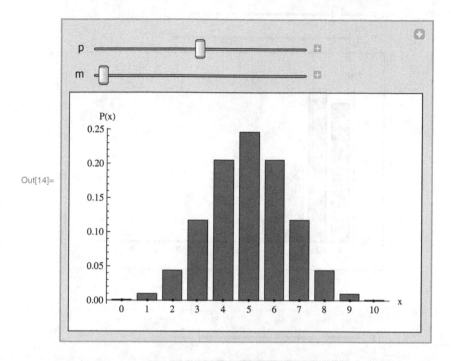

Moving the slider for parameter m increases the number of bars in our chart, but seemingly has a limited influence on its shape. When moving the slider for parameter p however, we observe a significant change in the shape of the distribution, moving from a positive skew as p approaches zero, through apparent symmetry around $p = 1/2$, to a negative skew as p approaches 1. This behaviour is easily understood. When p is close to 1, successful trials are far more likely than unsuccessful trials and *vice versa* when p is close to zero. Accordingly, we expect that $P(x)$ will exhibit a maximum close to $m p$.

Out[14]=

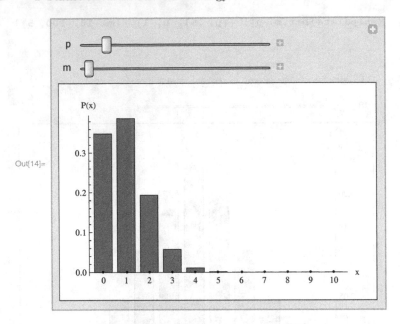

Out[14]=

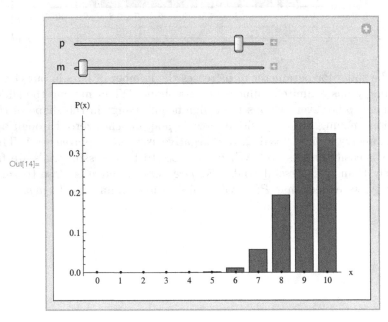

We quantify the asymmetry of the distribution by its *skewness*.

$$Sk(x) = \frac{\sum (x_i - \bar{x})^3 \, P(x)}{\sigma^3(x)} \tag{2.16}$$

such that for $Sk(x) > 0$ the distribution exhibits a longer tail to the right than to the left and *vice versa* for $Sk(x) < 0$; when $Sk = 0$ the distribution is symmetrical about the mean. For our binomial distribution, we have

In[15]:= **Skewness[BinomialDistribution[m, p]]**

Out[15]= $\dfrac{1 - 2\,p}{\sqrt{m\,(1 - p)\,p}}$

such that the distribution exhibits symmetry when $p = 1/2$, has a positive skew when $p < 1/2$ and a negative skew when $p > 1/2$. We observe also that the influence of p on the magnitude of skewness, and hence on the length of the tails of the distribution, diminishes as m increases. We observe behaviour consistent with this if we return to the bar chart that we generated with **Manipulate** and move the sliders to vary p and m.

2.3.3 Poisson Distribution

The Poisson distribution has probability function

$$P(x) = \frac{e^{-\bar{x}} \, \bar{x}^x}{x!} \qquad \text{for } x = 0, 1, 2, 3, \ldots \tag{2.17}$$

where $\bar{x}$ is the mean of the random variable x. It arises as a limiting case of the binomial distribution when the number of independent trials m is large and the probability of success p is small such that $m\,p = \bar{x}$.

The binomial coefficient, and hence the probability distribution function for the binomial distribution can be expressed in terms of the Euler gamma function, **Gamma**:

In[16]:= **FunctionExpand[PDF[BinomialDistribution[m, p], x]]**

Out[16]= $\dfrac{(1 - p)^{m-x}\,p^x\,\mathrm{Gamma}[1 + m]}{\mathrm{Gamma}[1 + m - x]\,\mathrm{Gamma}[1 + x]}$

The Euler gamma function is an example of a 'special function' that we shall encounter in several contexts as we develop our models. It satisfies,

$$\Gamma(z) = \int_0^\infty t^{z-1}\, e^{-t} \, dt \; . \tag{2.18}$$

The arguments to **Gamma** can take any value, but in the special case of integer arguments, the gamma function returns the factorial, such that

$$\Gamma(z+1) = z! \qquad \text{for } z = 0, 1, 2, 3 \ldots$$

The *Mathematica* command **FullSimplify** is able to reduce many expressions containing special functions to simpler forms, particularly if the second argument provides some assumptions, *e.g.*

In[17]:= **FullSimplify[%, {x, m} ∈ Integers && 0 ≤ x < m]**

Out[17]= $\dfrac{(1-p)^{m-x} \, p^x \, m!}{(m-x)! \; x!}$

Note that in the above example, we use the symbol ∈, input as $\boxed{\text{Esc}}\,$ **el** $\,\boxed{\text{Esc}}$, to specify that the variables x and m are elements of the domain **Integers**; the inequality ≤ is input using '**<=**'.

We have noted that the Poisson distribution arises from the binomial distribution when $m\,p = \bar{x}$ and for large m, so we substitute $\bar{x}/m$ for p and take the limit as $m \to \infty$. To perform the substitution, we use the command **ReplaceAll**, which we input as '**/.**' To input the arrow for the limit, we use '**->**', so we do,

In[18]:= **% /. p → (xbar / m)**
      Limit[%, m → ∞]

Out[18]= $\dfrac{\left(\frac{\text{xbar}}{m}\right)^x \left(1 - \frac{\text{xbar}}{m}\right)^{m-x} m!}{(m-x)! \; x!}$

Out[19]= $\dfrac{e^{-\text{xbar}} \, \text{xbar}^x}{x!}$

This last expression is the probability function for the Poisson distribution as given by Equation 2.17 and we note that the probability of observing a given integer x depends only on the expected value $\bar{x}$. We can call this probability function directly using

In[20]:= **PDF[PoissonDistribution[xbar], x]**

Out[20]= $\dfrac{e^{-\text{xbar}} \, \text{xbar}^x}{x!}$

and we note that the mean and variance of the Poisson distribution are equal:

In[21]:= **Mean[PoissonDistribution[xbar]]**
         Variance[PoissonDistribution[xbar]]

Out[21]= **xbar**

Out[22]= **xbar**

It follows that the coefficient of variation is $1/\sqrt{\bar{x}}$.

Many physical phenomena are described rather well by the Poisson distribution [53] and it is often considered to be the standard model for random processes. We shall use the distribution several times in the models we derive in the following chapters and will consider it to model pure random processes. Thus, if we partitioned the random networks shown in Figure 1.3 into say 10×10 square regions, we expect the number of fibre centres occurring within these regions to be distributed according to the Poisson distribution and so to have variance equal to the mean. The expected number of fibre centres in such regions will be the same for the disperse and clumped networks, but the variance of the number of fibre centres within regions would be less than the mean for the disperse cases, and greater than the mean in the clumped cases.

2.4 Common Probability Density Functions

So far, we have considered some of the more common statistical distributions that may be used to characterise discrete random variables. The functions that we have studied give the probability of a given outcome, say x, such that the probability $0 \le P(x) \le 1$. In Section 2.2 we observed that many random variables are not discrete, but are continuous. A property of a continuous random variable is that the probability of it having a given value x is infinitesimal, though the probability that x lies in a given interval is finite and lies between 0 and 1. The mathematical functions used to describe distributions of continuous variables are called probability *density* functions, whereas those describing the distributions of discrete random variables are called probability functions, or probability *distribution* functions. If we denote the probability density function of a continuous random variable x, $f(x)$, then the probability that x lies in the range $x_1 \le x < x_2$ is

$$P(x_1 \le x < x_2) = \int_{x_1}^{x_2} f(x) \, \mathrm{d}x \, . \tag{2.19}$$

If x is defined in the domain $x_{\min} \le x < x_{\max}$, then

$$\int_{x_{\min}}^{x_{\max}} f(x) \, \mathrm{d}x = 1 \, . \tag{2.20}$$

The probability that $x < X$ is called the cumulative distribution function. It is given by

$$F(X) = \int_{x_{\min}}^{X} f(x)\,\mathrm{d}x \ . \tag{2.21}$$

The mean is given by

$$\bar{x} = \int_{x_{\min}}^{x_{\max}} x\,f(x)\,\mathrm{d}x \ , \tag{2.22}$$

and the variance is given by

$$\sigma^2(x) = \int_{x_{\min}}^{x_{\max}} (x - \bar{x})^2\,f(x)\,\mathrm{d}x \ . \tag{2.23}$$

It is instructive to compare Equations 2.22 and 2.23 with Equations 2.8 and 2.9, respectively. We proceed by considering some common probability density functions encountered frequently in subsequent chapters.

2.4.1 Uniform Distribution

As expected, the uniform distribution is the continuous analogue of the discrete uniform distribution, which we considered in Section 2.2.1. Thus, whereas previously for the discrete random variable $x_{\min} \le x \le x_{\max}$ we required

$$\sum_{i=x_{\min}}^{x_{\max}} P(x) = 1 \ ,$$

for the continuous random variable distributed uniformly in the same domain we require,

$$\int_{x_{\min}}^{x_{\max}} f(x)\,\mathrm{d}x = 1 \ .$$

Accordingly, the uniform distribution has probability density given by

$$f(x) = \begin{cases} \dfrac{1}{x_{\max} - x_{\min}} & \text{if } x_{\min} \le x \le x_{\max} \\ 0 & \text{otherwise.} \end{cases} \tag{2.24}$$

We obtain this probability density function in *Mathematica* using

In[23]:= **PDF[UniformDistribution[{xmin, xmax}], x]**

Out[23]= $\left\{ \dfrac{1}{\text{xmax-xmin}} \quad \text{xmin} \le \text{x} \le \text{xmax} \right.$

The mean and variance are

```
In[24]:= Mean[UniformDistribution[{xmin, xmax}]]
         Variance[UniformDistribution[{xmin, xmax}]]
```

$$\text{Out[24]= } \frac{\text{xmax} + \text{xmin}}{2}$$

$$\text{Out[25]= } \frac{1}{12} \, (\text{xmax} - \text{xmin})^2$$

In the following plots of the probability density function we use the option
Exclusions -> None to connect the discontinuities in the function.

```
In[26]:= GraphicsGrid[
         {{Plot[PDF[UniformDistribution[{1, 2}], x],
            {x, 0, 3}, Exclusions → None],
          Plot[PDF[UniformDistribution[{1, 1.5}], x],
            {x, 0, 3}, Exclusions → None]}}]
```

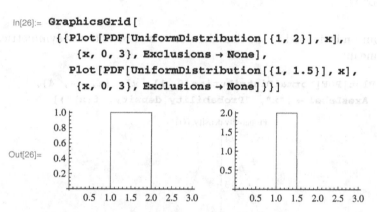

The plot on the right is generated for the uniformly distributed random variable in a domain where $(x_{\max} - x_{\min}) < 1$ such that $f(x) > 1$. This is an important feature of probability density functions, whereas probabilities must be between 0 and 1, probability densities can exceed 1.

2.4.2 Normal Distribution

Most of us have encountered the classical bell-shaped normal, or Gaussian, distribution. It describes the distribution of data arising in many physical and biological contexts very well. The normal distribution is fully defined by its mean, μ and variance σ^2 and has probability density given by

$$f(x) = \frac{1}{\sqrt{2\pi}\,\sigma} \, e^{-\frac{(x-\mu)^2}{2\sigma^2}} . \tag{2.25}$$

We call the probability density function, mean and variance in *Mathematica* in the usual way:

In[27]:= **PDF[NormalDistribution[μ, σ], x]**
Mean[NormalDistribution[μ, σ]]
Variance[NormalDistribution[μ, σ]]

Out[27]=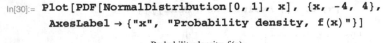

Out[28]= μ

Out[29]= σ^2

The distribution is defined in the domain $-\infty < x < \infty$ and it is symmetrical about the mean:

In[30]:= **Plot[PDF[NormalDistribution[0, 1], x], {x, -4, 4},**
AxesLabel → {"x", "Probability density, f(x)"}]

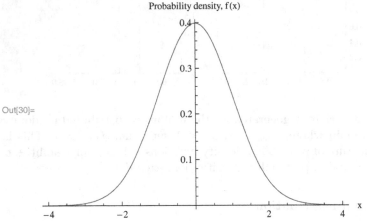

Out[30]=

We have already noted that the normal distribution is often found to describe distributions encountered in a wide variety of contexts. This very convenient property of many random variables can be attributed to the central limit theorem. Here we state the central limit theorem in simple terms following Chatfield [14]; detailed discussion and proof of the theorem are given by, *e.g.* Papoulis [119], pp. 278–284.

Consider the random variable

$$x = \frac{(x_1 + x_2 + \dots x_n)}{n}$$

where the x_i are drawn from independent and identical distributions with mean μ and variance σ^2. The central limit theorem states that the distribution of x is approximately normal with mean μ and variance σ^2/n and that as n increases, so does the quality of the approximation. Importantly, the source distribution of the x_i does not need to be specified; the central limit theorem applies for any source distribution.

The following *Mathematica* code illustrates the central limit theorem by plotting the histogram of 50,000 independent x where the source distribution is a uniform distribution on the interval $\{-1, 1\}$ which has mean $\mu = 0$ and variance $1/3$. With each histogram, we plot the probability density function of the normal distribution with the same mean and variance $1/(3\,n)$.

```
In[31]:= σ = StandardDeviation[UniformDistribution[{-1, 1}]];
        Needs["Histograms`"]
        H[n_] :=
          Show[Histogram[Mean[RandomReal[{-1, 1}, {n, 50 000}]],
            HistogramCategories → 50, HistogramScale → 1],
            Plot[PDF[NormalDistribution[0, σ/√n], x],
            {x, -1, 1}, PlotStyle → Thickness[.015]]]
        GraphicsArray[{{H[1], H[2]}, {H[3], H[10]}}]
        σ = .
```

Out[34]=

The graphic in the top left shows the histogram for our source uniform distribution, which is manifestly not well approximated by the normal distribution. On the top right, we see the histogram for the case when $n = 2$. Here, each of our values of x is the mean of two values drawn randomly from the uniform distribution; the resultant histogram exhibits a symmetrical triangular distribution. On the bottom left, $n = 3$ and we can see quite clearly that the approximation of the histogram to the normal distribution is improving. In the final case considered here, as shown on the bottom right, $n = 10$ and the approximation is very good. Of course, our example is rather static, but it is simple to generate a **Manipulate** object to investigate the influence of n interactively.

  Manipulate[H[m], {m, 1, 20, 1}]

The code that we have used to demonstrate the central limit theorem introduces a few *Mathematica* commands. **Histogram** operates on a one-dimensional list to count the occurrences of data in intervals, the width of which are determined by the maximum and minimum values occurring in the list and the number of categories; we have specified this using the option **HistogramCategories**, but if this is not specified, the number of categories are chosen automatically. The option **HistogramScale -> 1** scales the heights of the bars so that the area under the histogram is 1, allowing direct comparison with a plot of a probability density function, as we have done here. The uniformly distributed random data in the interval {-1, 1} are generated using **RandomReal** as opposed to the command **RandomInteger** used earlier. Here we have specified an interval, though if we input **RandomReal[ ]** then we would obtain uniformly distributed random numbers between 0 and 1. We have encountered the command **Mean** earlier, but here we take advantage of the way that it operates on a list. In our example, **Mean** operates on a list consisting of n sublists, each of length 50,000, and computes the mean of the ith value of each of these sublists to generate a new list of length 50,000, where each element represents a value of x. The manner in which **Mean** operates on a list of sublists is perhaps best understood by examining the following example:

In[36]:= **Mean[{{a, b, c}, {d, e, f}, {g, h, i}}]**

Out[36]= $\left\{ \frac{1}{3} (a + d + g), \ \frac{1}{3} (b + e + h), \ \frac{1}{3} (c + f + i) \right\}$

We will make considerable use of *Mathematica*'s advanced list handling capabilities when applying Monte Carlo techniques to the modelling of fibre networks, and by understanding precisely how they work, we are able to optimise our code, reducing evaluation times considerably.

2.4.3 Lognormal Distribution

The random variable x has a lognormal distribution if the random variable $y = \log(x)$ is normally distributed with mean μ and variance σ^2. The probability density for the lognormal distribution is given by

$$f(x) = \begin{cases} \frac{1}{\sqrt{2\pi}\,\sigma\,x} \, e^{-\frac{(\log(x) - \mu)^2}{2\sigma^2}} & \text{if } x \geq 0 \\ 0 & \text{otherwise.} \end{cases} \qquad (2.26)$$

It is obtained using one of the variable transform formulae. We shall encounter variable transform formulae for the sums and products of independent random variables repeatedly in subsequent chapters. The simplest variable transform formula allows us to obtain the probability density of $y = g(x)$ if the probability density of x is $f(x)$. We state it as follows:

$$f(y) = f(g(x)) \left| \frac{dy}{dx} \right| . \tag{2.27}$$

The second term on the right-hand side of Equation 2.27 is called the Jacobian and takes account of the change in the domain of the random variable such that the resultant probability density integrates to 1 over its domain. Note that to perform this variable transformation, we require the relationship between y and x to be one-to-one, *i.e.* each value of y is associated with only one value of x. We are interested in the variable $y = \log(x)$ which is indeed one-to-one so we perform the change of variable by doing,

```
In[37]:= y = Log[x];
         PDF[NormalDistribution[μ, σ], y] D[y, x]
         y =.
```

$$Out[38]= \frac{e^{-\frac{(-\mu+\text{Log}[x])^2}{2\sigma^2}}}{\sqrt{2\pi}\, x\, \sigma}$$

Which is the probability density function for the lognormal distribution as given by Equation 2.26. We call this probability density in *Mathematica* using

```
PDF[LogNormalDistribution[μ, σ], x]
```

The probability density is defined in the domain $0 \leq x \leq \infty$ and it exhibits a positive skew.

```
In[40]:= Plot[PDF[LogNormalDistribution[1, 1], x], {x, 0, 20},
         AxesLabel → {"x", "Probability density, f(x)"}]
```

Probability density, f(x)

Out[40]=

We obtain the mean and variance in the usual way and from these can calculate the coefficient of variation using **PowerExpand** in conjunction with **FullSimplify**:

```
In[41]:= Mean[LogNormalDistribution[μ, σ]]
        Variance[LogNormalDistribution[μ, σ]]
        FullSimplify[PowerExpand[Sqrt[%] / %%]]
```

$$\text{Out[41]}= \; e^{\mu + \frac{\sigma^2}{2}}$$

$$\text{Out[42]}= \; e^{2\mu + \sigma^2}\left(-1 + e^{\sigma^2}\right)$$

$$\text{Out[43]}= \; \sqrt{-1 + e^{\sigma^2}}$$

We observe then that the mean and variance of the lognormal distribution are defined in terms of those of the normal distribution, μ and σ^2, respectively, and that the coefficient of variation depends only on σ.

We have just seen that the normal distribution arises as a consequence of the central limit theorem when considering the sum of several x_i drawn from independent and identical distributions. The central limit theorem provides also the explanation for the occurrence of the lognormal distribution. Consider the product of n independent positive random variables x_i:

$$x = x_1 x_2 \ldots x_n$$

The random variable $y = \log(x)$ is given by

$$y = \log(x) = \log(x_1) + \log(x_2) + \ldots + \log(x_n)$$

Since random variables arising as the sum of independent random variables exhibit a normal distribution, we may state that the random variable $y = \log(x)$ exhibits a normal distribution and thus the random variable x exhibits a lognormal distribution.

2.4.4 Exponential distribution

The random variable x with mean $\bar{x}$ is said to have an exponential distribution if its probability density is given by

$$f(x) = \begin{cases} \frac{1}{\bar{x}} e^{-\frac{x}{\bar{x}}} & \text{if } x \geq 0 \\ 0 & \text{otherwise.} \end{cases} \tag{2.28}$$

In Section 3.3.2 we will show that the exponential distribution arises as the distribution of intervals between random events as described by the Poisson distribution. For now, we note that the probability density given by Equation 2.28 is a weighted form of the probability of zero events in a point Poisson process where the discrete random variable is the occurrence of events in an interval.

We obtain the exponential probability density and its mean and variance in *Mathematica* using,

```
In[44]:= PDF[ExponentialDistribution[1/xbar], x]
         Mean[ExponentialDistribution[1/xbar]]
         Variance[ExponentialDistribution[1/xbar]]
```

$$Out[44]= \frac{e^{-\frac{x}{xbar}}}{xbar}$$

$$Out[45]= xbar$$

$$Out[46]= xbar^2$$

Note that since the variance of the distribution is the square of the mean, the exponential distribution has constant coefficient of variation $cv(x) = 1$. Accordingly, the shape of the distribution is unaltered by the mean which acts as a scaling parameter.

```
In[47]:= Plot[PDF[ExponentialDistribution[1], x], {x, 0, 5},
         AxesLabel → {"x", "Probability density, f(x)"},
         PlotRange → All]
```

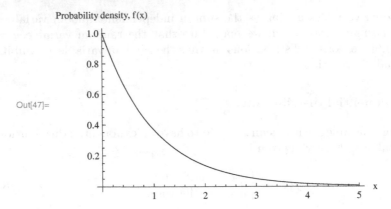

2.4.5 Gamma Distribution

The random variable x is said to have a gamma distribution if its probability density is given by

$$f(x) = \left(\frac{\alpha}{\beta}\right)^{\alpha} \frac{x^{\alpha-1}}{\Gamma(\alpha)} e^{-\alpha x/\beta} , \qquad (2.29)$$

where the term $\Gamma(\alpha)$ represents the Euler gamma function, which we encountered on page 35. We obtain the probability density, mean, variance and coefficient of variation as previously:

In[48]:= **PDF[GammaDistribution[α, β / α], x]**
Mean[GammaDistribution[α, β / α]]
Variance[GammaDistribution[α, β / α]]
PowerExpand$\left[\sqrt{\text{%}} \Big/ \text{%%}\right]$

Out[48]= $\dfrac{e^{-\frac{x\alpha}{\beta}} x^{-1+\alpha} \left(\frac{\beta}{\alpha}\right)^{-\alpha}}{\text{Gamma}[\alpha]}$

Out[49]= β

Out[50]= $\dfrac{\beta^2}{\alpha}$

Out[51]= $\dfrac{1}{\sqrt{\alpha}}$

So, in the form given in Equation 2.29, the distribution has mean $\bar{x} = \beta$ and coefficient of variation $cv(x) = 1/\sqrt{\alpha}$.

Although the probability density for the gamma distribution is convention-
ally expressed in terms of parameters α and β, it can be expressed equally
well in terms of the mean and coefficient of variation:

In[52]:= **PDF$\left[\text{GammaDistribution}\left[1/\text{cv}^2,\ \text{xbar cv}^2\right],\ \text{x}\right]$**
Mean$\left[\text{GammaDistribution}\left[1/\text{cv}^2,\ \text{xbar cv}^2\right]\right]$
Variance$\left[\text{GammaDistribution}\left[1/\text{cv}^2,\ \text{xbar cv}^2\right]\right]$
PowerExpand$\left[\sqrt{\%}\ /\ \%\%\right]$

Out[52]= $\dfrac{e^{-\frac{x}{cv^2\ xbar}}\ x^{-1+\frac{1}{cv^2}}\left(cv^2\ xbar\right)^{-\frac{1}{cv^2}}}{\text{Gamma}\left[\frac{1}{cv^2}\right]}$

Out[53]= xbar

Out[54]= $cv^2\ xbar^2$

Out[55]= cv

The mean $\bar{x} = \beta$ acts as a scaling parameter for the gamma distribution and
the coefficient of variation $cv(x) = 1/\sqrt{\alpha}$ controls its shape. We illustrate
the influence of the coefficient of variation on the shape of the distribution
by comparing plots of the probability density for distributions with unit mean:

In[56]:= **PlotPDF[cv_] :=**
Plot$\left[\text{PDF}\left[\text{GammaDistribution}\left[1/\text{cv}^2,\ \text{cv}^2\right],\ \text{x}\right]$,
{x, 0, 5}, PlotRange → {All, {0, 2}},
PlotLabel → Row[{"cv =", cv}]]
GraphicsArray[{{PlotPDF[0.25], PlotPDF[0.5]},
{PlotPDF[1], PlotPDF[2]}}]

Out[57]=
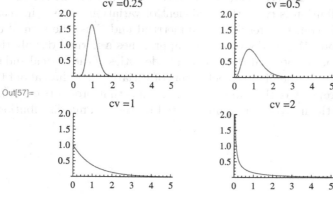

We observe that the distribution exhibits a positive skew and this increases with the coefficient of variation:

In[58]:= **PowerExpand[Skewness[GammaDistribution[1/cv², cv²]]]**

Out[58]= 2 cv

Note that the distribution exhibits a maximum only when $cv(x) < 1$ and decreases monotonically otherwise.

The gamma distribution includes the exponential distribution as a special case when $\alpha = 1$:

In[59]:= **TrueQ[GammaDistribution[1, β] ==**
        ExponentialDistribution[1/β]]

Out[59]= True

Recall that the coefficient of variation of the exponential distribution is 1 and that of the gamma distribution is $1/\sqrt{\alpha}$. This means that if a random variable has a gamma or exponential distribution, then processes that change the mean will change the standard deviation proportionately and a plot of the standard deviation against the mean will be linear with gradient $cv(x)$. Hwang and Hu [69] provide a proof that for independent positive random variables $x_1, x_2, \ldots, x_n$ with a common continuous probability density function, this property of the sample mean $\bar{x}$ and coefficient of variation $cv(x)$ being independent is equivalent to the x_i being drawn from a gamma distribution. Such linearity is common in experimental data characterising the void structure of fibrous materials and Johnston [72, 73] proposes that the gamma distribution describes the pore size distribution of stochastic porous materials in general.

It turns out that the sum of n independent exponential random variables is a gamma distribution. This is consistent with our remarks concerning the central limit theorem on page 41. The central limit theorem gives $cv(x, n) = 1/\sqrt{n}$ such that as n increases, $cv(x, n)$ decreases. Recall that the skewness of the gamma distribution is twice the coefficient of variation, so as n increases the distribution becomes increasingly symmetrical and thus, the sum of our independent exponential random variables approaches a normal distribution. As a rule of thumb, we note that the probability densities of the normal and the gamma distribution are similar for coefficients of variation less than about 0.2. Another useful property of the gamma distribution is that products of gamma distributions are themselves well approximated by the gamma distribution.

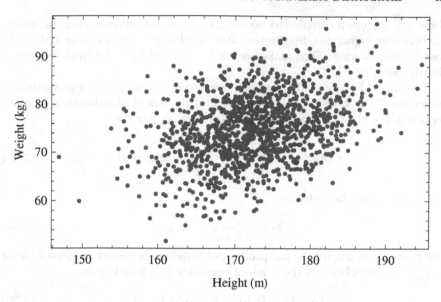

Figure 2.1. Simulated data for typical weight and height distributions of men

2.5 Multivariate Distributions

So far we have considered the distributions of single random variables and those arising from combinations of several independent distributions. Very often, we need to consider distributions of *dependent* random variables. For example, if we recorded the heights and weights of students at a university we would expect both variables to exhibit a normal distribution. Equally, we might expect taller students to be heavier than shorter students, but we would not be surprised to find tall students who were lighter than average or short students who were heavier. Thus, we would expect a scatter plot of weight against height for male students to look something like Figure 2.1.

Our example considers two random variables, height and weight, which exhibit a degree of interdependence. As such, we classify the distribution as being bivariate. Since the distributions of height and weight both exhibit a normal distribution, then the data in Figure 2.1 exhibit a bivariate normal distribution. We shall consider this distribution in more detail shortly, but first we note that bivariate distributions often occur in stochastic fibrous materials and in the fibres from which they are formed. For example, the length and diameter of wool fibres [94], man-made mineral fibres [67, 150] and wood-pulp fibres [85] are often distributed according to bivariate distributions including the bivariate lognormal distribution and the distributions of mass and thickness in near-random fibre mats are well described by the bivariate normal distribution [36, 37].

To handle bivariate distributions, we require some additional statistical descriptors. The *covariance* is a measure of the association between random variables. For the random variables x and y it is given by

$$Cov(x, y) = \frac{1}{n-1} \sum_{i=1}^{n} (x_i - \bar{x})(y_i - \bar{y}) , \qquad (2.30)$$

and it is easy to show that

$$Cov(x, y) = \overline{xy} - \bar{x}\,\bar{y} . \qquad (2.31)$$

The units of covariance are the product of those of the constituent variables x and y. The covariance of the random variables $k_x\,x$ and $k_y\,y$ is

$$Cov(k_x\,x, k_y\,y) = k_x\,k_y\,Cov(x, y) , \qquad (2.32)$$

and the variance of the sum of the random variables x and y is

$$\sigma^2(x+y) = \sigma^2(x) + 2\,Cov(x, y) + \sigma^2(y) . \qquad (2.33)$$

Often, we seek to standardize the covariance so that it is dimensionless. It is convenient to do this by dividing $(x_i - \bar{x})$ and $(y_i - \bar{y})$ by their standard deviations $\sigma(x)$ and $\sigma(y)$, respectively. The resultant expression gives the *correlation* between the variables, $-1 \leq \rho \leq 1$:

$$\rho = \frac{1}{n-1} \sum_{i=1}^{n} \frac{(x_i - \bar{x})}{\sigma(x)} \frac{(y_i - \bar{y})}{\sigma y} \qquad (2.34)$$

$$= \frac{Cov(x, y)}{\sigma(x)\,\sigma(y)} , \qquad (2.35)$$

where ρ is called the correlation coefficient.

Negative covariance and correlation tells us that as the value of one variable increases, so that of the other decreases; positive correlation tells us that an increase in one variable is associated with an increase in the other; if the variables are independent, then the correlation and covariance are zero[2]. The size of the correlation coefficient tells us the degree of scatter in a plot such as Figure 2.1, so can be interpreted as providing a measure of the range of spread of x and y about a regression line. Often the square of the correlation coefficient, is reported for experimental data; this parameter, typically denoted r^2, is called the coefficient of determination.

---

[2] The converse is not necessarily true [119].

2.5.1 Bivariate Normal Distribution

As we might expect given our earlier discussion, the bivariate distribution encountered most frequently is the bivariate normal, or Gaussian, distribution. The bivariate normal distribution arises when the random variables x and y both exhibit a normal distribution and are correlated. This means that if we select any x_i from a bivariate normal distribution, then we expect the range of y_i that can be associated with it to be constrained to an extent depending on the correlation ρ. We state then that x and y have *joint* probability density.

For the bivariate normal distribution, the joint probability density of x and y with correlation coefficient ρ is

$$f(x,y) = \frac{1}{2\pi\,\sigma(x)\,\sigma(y)\,\sqrt{1-\rho^2}}\,e^{-\frac{z}{2(1-\rho^2)}} \qquad (2.36)$$

where

$$z = \frac{(x-\bar{x})^2}{\sigma^2(x)} - \frac{2\rho\,(x-\bar{x})(y-\bar{y})}{\sigma(x)\,\sigma(y)} + \frac{(y-\bar{y})^2}{\sigma^2(y)}$$

In *Mathematica* we obtain the joint probability density function in this form using the command **MultinormalDistribution** which is in the **MultivariateStatistics** package:

```
In[60]:= Needs["MultivariateStatistics`"]
         FullSimplify[PDF[MultinormalDistribution[{xbar, ybar},
           {{σx², ρ σx σy}, {ρ σx σy, σy²}}], {x, y}],
         {σx, σy} > 0 && {σx, σy} ∈ Reals]
```

$$\text{Out[61]=}\quad \frac{\mathbb{e}^{\frac{(y-ybar)^2\,\sigma x^2 - 2\,(x-xbar)\,(y-ybar)\,\rho\,\sigma x\,\sigma y + (x-xbar)^2\,\sigma y^2}{2\,(-1+\rho^2)\,\sigma x^2\,\sigma y^2}}}{2\,\pi\,\sqrt{1-\rho^2}\,\text{Abs}[\sigma x]\,\text{Abs}[\sigma y]}$$

Note that we are considering only the bivariate normal distribution here, but the command **MultinormalDistribution** will handle trivariate and higher orders of multivariate normal distributions. Accordingly, the second argument for an n-variate distribution is input as an $n \times n$ array known as the covariance matrix where the elements are the covariances of the variables with each other. From the definition of covariance $Cov(x,x) = \sigma^2(x)$, so the first and last terms are simply the variances of x and y and the matrix is symmetrical. *Mathematica* returns the mean and variance of the bivariate normal distribution as two element lists, and the covariance as a matrix in the same form as it was input:

```
In[62]:= Mean[MultinormalDistribution[
            {xbar, ybar}, {{σx², ρ σx σy}, {ρ σx σy, σy²}}]]
         Variance[MultinormalDistribution[{xbar, ybar},
            {{σx², ρ σx σy}, {ρ σx σy, σy²}}]]
         Covariance[MultinormalDistribution[
            {xbar, ybar}, {{σx², ρ σx σy}, {ρ σx σy, σy²}}]]
```

Out[62]= {xbar, ybar}

Out[63]= {σx², σy²}

Out[64]= {{σx², ρ σx σy}, {ρ σx σy, σy²}}

The joint probability density is represented by a surface. Here we generate this for a bivariate normal distribution with mean $\bar{x} = \bar{y} = 0$, variances $\sigma^2(x) = 1$, $\sigma^2(y) = \frac{1}{4}$ and correlation $\rho = 0.8$. Note the use of the command **Clear** to unset several variables; this command may also be used to clear functions which have been previously defined.

```
In[65]:= σx = 1;
         σy = 1 / 2;
         ρ = 0.8;
         Plot3D[PDF[MultinormalDistribution[{0, 0},
             {{σx², ρ σx σy}, {ρ σx σy, σy²}}], {x, y}],
           {x, -4, 4}, {y, -2, 2}, PlotRange → All,
           AxesLabel → {"x", "y", "f(x,y)"}]
         Clear[σx, σy, ρ]
```

Out[68]=

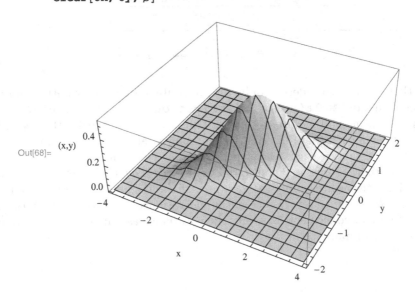

The fraction of the distribution which lies in the interval $(x + \Delta x, y + \Delta y)$ is

$$P(x, y) = \int_x^{x+\Delta x} \int_y^{y+\Delta y} f(x, y) \, dy \, dx \, , \tag{2.37}$$

and the joint probability density integrates to unity (*cf.* Equation 2.20):

$$\int_{-\infty}^{\infty} \int_{-\infty}^{\infty} f(x, y) \, dy \, dx = 1 \, . \tag{2.38}$$

3

Planar Poisson Point and Line Processes

3.1 Introduction

Many fibrous materials, such as those shown in Figure 1.1 are very close to being two-dimensional because their dimensions perpendicular to the plane are very small relative to in-plane dimensions. For example, the paper on which this book is printed has thickness of order 100 µm, so the in-plane dimensions of the smallest sample likely to have any practical value will be far greater than this. We classify such materials as being *near-planar* and will consider the structural properties of these in detail in Section 4.3. In order to understand the structure of near-planar fibrous materials, and to provide ourselves with the analytic tools required to model them, we must first consider the properties of idealised structures which occur only in two dimensions, *i.e.* networks of model fibres with thickness zero, which we classify as planar networks.

3.2 Point Poisson Processes

We shall consider first pure random processes where each event is independent of all others. Such processes are modelled using the Poisson distribution, which we encountered on page 35. Recall that the Poisson distribution gives the probability of discrete random variables and the distribution is fully characterised by the mean. Our interest is random networks of fibres in the plane; initially, the number of fibres per unit area of the network might seem an attractive choice of the discrete random variable to model. Referring to Figure 3.1, we observe that this is not a discrete random variable since fibres are extended objects and may therefore exist in more than one area. The centre of each fibre however – identified by the points within the fibres in Figure 3.1 – represents a single point in the plane, and thus can exist only in one area. Thus, the process we consider is the occurrence of fibre centres in the plane where the number of fibre centres in an area has a Poisson distribution.

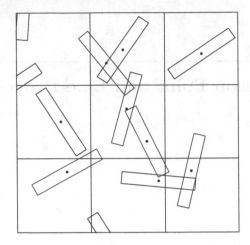

Figure 3.1. Examples of fibres placed randomly within contiguous square inspection areas. One fibre can exists in several inspection areas, whereas fibre centres can exist in only one inspection area

3.2.1 Clustering

We have already commented that random processes are not uniform, but exhibit clustering. We can easily visualise this clustering by plotting points with coordinates drawn from a uniform distribution. In the example that follows we use the command **RandomReal** to generate such coordinates $0 \leq x, y \leq 10$. Throughout, we will refer to data generated in this way as being random, though they are more precisely referred to as pseudorandom numbers. *Mathematica* can generate pseudorandom numbers in a variety of ways that can be specified as options to the command **RandomReal**. The default algorithm used by *Mathematica* involves the use of a cellular automaton and provides pseudorandom numbers of sufficient quality that we may consider them to be random.

We generate a plot of 1,000 random x, y pairs within a square region of side 10 as follows:

```
In[1]:= SeedRandom[1]
        NCentres = 1000;
        xyPairs = RandomReal[{0, 10}, {NCentres, 2}];
        ListPlot[xyPairs, PlotRange → {{0, 10}, {0, 10}},
          AspectRatio → Automatic, Frame → True]
        Clear[NCentres]
```

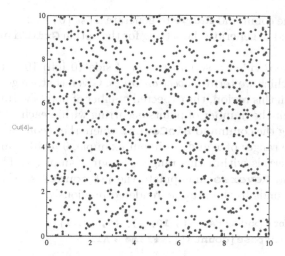

where the command **SeedRandom[1]** is included to allow precisely the same data to be generated for repeated evaluations of the same code including different installations. As we mentioned on page 4, it is immediately clear that the points are not uniformly distributed in the plane, but exhibit clustering. We can quantify this clustering by partitioning the region into contiguous square zones and counting the number of fibre centres within each zone. We illustrate this process by adding gridlines to our plot of x, y pairs:

```
In[6]:= grids = Range[0, 10];
       ListPlot[xyPairs, PlotRange → {{0, 10}, {0, 10}},
         AspectRatio → Automatic, Frame → True,
         GridLines → {grids, grids}]
```

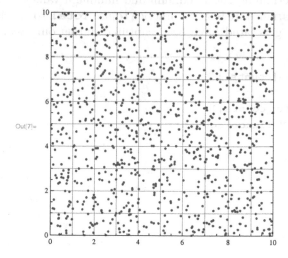

The first line of code uses the command **Range** to generate a list $\{0,1,2\ldots,$ $10\}$ that is subsequently used to generate the setting for the option **GridLines** when calling **ListPlot**.

In our example, we have divided a square region of side 10 into 10×10 contiguous square regions which we term *inspection zones*. Given that we generated 1,000 x, y pairs within this region, the expected number of points within a unit square is 10. We count the number of points occurring within each unit square inspection zone using the command **BinCounts** where the second and third arguments specify the ranges of x and y – in our example, 0 to 10 – and the size of the required intervals within this range – in our example, 1. The commands **Transpose** and **Reverse** manipulate the list **counts** so that the output of **Grid** corresponds to the graphic rendered previously.

```
In[8]:= counts = BinCounts[xyPairs, {0, 10, 1}, {0, 10, 1}];
        Grid[Reverse[Transpose[counts]], Frame → All]
```

Out[9]=

7	12	9	8	9	9	6	12	11	13
10	7	14	14	7	7	11	11	5	12
11	6	11	11	12	17	7	9	14	8
10	14	7	11	4	9	8	8	13	11
10	11	7	13	10	9	13	14	13	10
11	5	10	13	14	6	13	11	12	8
4	7	9	17	5	9	16	11	14	6
14	8	9	10	11	7	3	11	6	11
9	11	11	14	7	7	8	8	16	7
14	8	8	14	7	7	14	11	7	16

The parameter **counts** is a nested list, so we must flatten it to make it one dimensional before determining the maximum and minimum values and calculating the mean and variance. It is good practice to use **Total** to check that all the points initially generated have been captured during our treatment.

```
In[10]:= counts = Flatten[counts];
         Total[counts]
         {Min[counts], Max[counts]}
         {Mean[counts], Variance[counts]}
```

Out[11]= 1000

Out[12]= {3, 17}

$$\text{Out[13]}= \left\{10, \frac{320}{33}\right\}$$

We note that the variance is close to the mean, and recall that a property of the Poisson distribution is that the variance is equal to the mean. The command **Tally** returns a list where each entry is a sublist consisting of each value and its frequency of occurrence and the output is given in order of the first occurrence of each value. To obtain these frequencies in order of the value of the variable being counted, we use **Sort**:

In[14]:= **barcounts = Sort[Tally[counts]]**

Out[14]= {{3, 1}, {4, 2}, {5, 3}, {6, 5},
       {7, 15}, {8, 10}, {9, 10}, {10, 7}, {11, 18},
       {12, 5}, {13, 7}, {14, 12}, {16, 3}, {17, 2}}

This output provides the input and option specifications to generate a bar chart of the frequency of occurrence of x, y pairs within unit squares.

In[15]:= **Needs["BarCharts`"]**
       barcounts = Transpose[barcounts]
       BarChart[barcounts[[2]], BarLabels → barcounts[[1]]]
       Total[barcounts[[2]]]

Out[16]= {{3, 4, 5, 6, 7, 8, 9, 10, 11, 12, 13, 14, 16, 17},
       {1, 2, 3, 5, 15, 10, 10, 7, 18, 5, 7, 12, 3, 2}}

Out[17]=
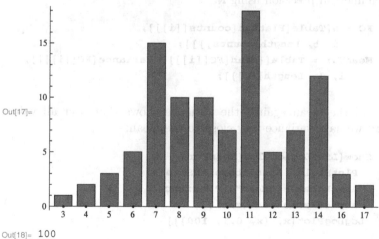

Out[18]= 100

Although the mean and variance of our data are similar, the bar chart we have generated does not resemble the probability function of the Poisson distribution, or that of any other of the probability functions that we considered in Chapter 2. We note however that our statistics have been derived by considering only a sample of size 100, which is certainly insufficient to fit a distribution to the frequency data. To obtain more reliable statistics characterising the distribution we must consider more points within our 10×10 square and the effect of dividing the square into a larger number of contiguous square inspection zones.

The first three lines of the following code correspond to those we used earlier, but generate 10,000 random x, y pairs. The final line nests the command **BinCounts** within a **Table** where the iterator **Nsquares** specifies that the 10×10 region is divided into **Nsquares** $\times$ **Nsquares** contiguous square zones. Thus, whereas in the last example we divided our 10×10 region into 10×10 contiguous square zones, here we divide it into **Nsquares** $\times$ **Nsquares** zones and allow **Nsquares** to take values between 10 and 100.

```
In[19]:= NCentres = 10 000;
        SeedRandom[1]
        xyPairs = RandomReal[{0, 10}, {NCentres, 2}];
        counts = Table[BinCounts[xyPairs, {0, 10, 10 / Nsquares},
            {0, 10, 10 / Nsquares}], {Nsquares, 10, 100}];
```

Before computing the mean and variance of counts in each of our sublists, we need to flatten each sublist; to speed up the computation, we convert integer values to numerical precision using **N**:

```
In[23]:= FC = N[Table[Flatten[counts[[i]]],
            {i, 1, Length[counts]}]];
        MeanVar = Table[{Mean[FC[[i]]], Variance[FC[[i]]]},
            {i, 1, Length[FC]}];
```

A plot of the mean against the variance shows that over all values of **Nsquares** we have variance very close to the mean:

```
In[25]:= Show[ListLogLogPlot[MeanVar,
            PlotStyle → AbsolutePointSize[4],
            AxesLabel → {"Mean", "Variance"},
            AspectRatio → Automatic],
        LogLogPlot[x, {x, 0.1, 100}]]
```

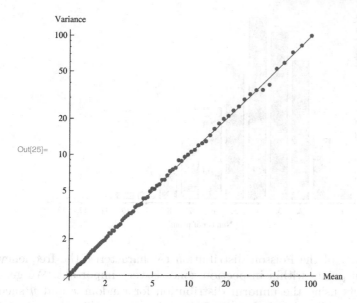

Naturally, as the number of inspection zones increases, so the expected number of points occurring within those zones decreases and we see that by computing our statistics for a larger number of inspection zones we reduce the amount of scatter of points around the line of unit gradient in a plot of the variance against the mean.

We now have much stronger data to suggest that the number of points occurring within inspection zones is distributed according to the Poisson distribution, with variance equal to the mean. Since we have examined a process of 10,000 points, when **Nsquares** = 50, the expected number of points in an inspection zone is 4. The data corresponding to **Nsquares** = 50 occur at position 41 in our list **FC**, so we compare a bar chart of the frequencies of counts in this sublist with those arising from a Poisson distribution with mean 4 and observe that they agree quite closely.

```
In[26]:= barcounts = Sort[Tally[FC[[41]]]];
         barcounts = Transpose[barcounts];
         frequencies = N[barcounts[[2]] / Length[FC[[41]]]];
         PoissonFrequencies = Table[{x + 1,
             PDF[PoissonDistribution[4], x]}, {x, 0, 13}];
         Show[BarChart[frequencies, BarLabels → barcounts[[1]]],
          ListLinePlot[PoissonFrequencies],
          Frame → {True, True, False, False},
          FrameLabel → {"Number of points", "Frequency"}]
```

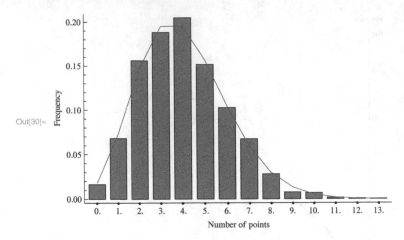

Out[30]=

The suitability of the Poisson distribution to characterise the frequency of occurrence of points within inspection zones is very important. We generated our points using the uniform distribution for random x and y such that they have equal likelihood of occupying any location within our region of interest. We have seen from plots of such points that these exhibit clustering and now we observe that this clustering is well characterised by the Poisson distribution.

> On page 4 we defined random fibre networks following Kallmes and Corte [74] and our definition stated that fibres were deposited independently of each other and had equal probability of landing at all points in the network. We state then that to precisely meet these criteria when modelling planar random fibre networks, we assume that fibre centres are distributed according to a point Poisson process in the plane.

It is instructive also to consider the number of points *per unit area* that occur within our inspection zones. This parameter is given by dividing each element within our list **FC** by the square of the side length of our square inspection zones:

In[31]:=

```
sides = 10 / Range[10, 100];
FCpua = FC / sides^2;
```

The expected number of points per unit area is, of course, unaffected by the size of our inspection zones. The variance is affected however, and decreases with the size of the inspection zone.

In[33]:= **Varpua =**
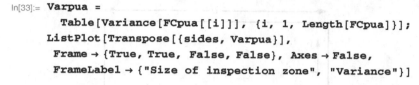

Table[Variance[FCpua[[i]]], {i, 1, Length[FCpua]}];
ListPlot[Transpose[{sides, Varpua}],
 Frame → {True, True, False, False}, Axes → False,
 FrameLabel → {"Size of inspection zone", "Variance"}]

Out[34]=
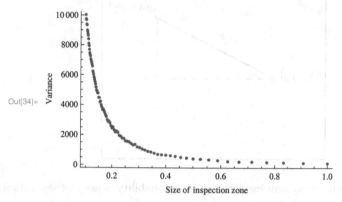

The observed decay of variance with zone size is very important and similar plots would, of course, be obtained if we were to plot the standard deviation or coefficient of variation against zone size. For our example, we could use the command **ListLogLogPlot** to demonstrate that the decay is logarithmic, though when we extend our analysis to consider fibres rather than points we shall see that this need not be the case and that the rate that variance decays with zone size reveals information about the texture of a fibre network and the sizes of the objects that contribute to the observed variability. At this stage it is sufficient to note that the observed variability depends upon the scale of inspection and thus this should be stated whenever the variance, standard deviation or coefficient of variation are reported. Correctly, the number of observations per unit area within each inspection zone is termed the *local average* number and varies from region to region; the number of observations per unit area in the region as a whole is termed the *global average*.

3.2.2 Separation of Pairs of Points

Given that point Poisson processes in the plane exhibit clustering, a natural property of interest related to the degree of clustering is the distribution of distances between pairs of points. Of course, for any given pair of points (x_1, y_1) and (x_2, y_2), we simply use Pythagoras' theorem. So, using the coordinate system given in Figure 3.2 the distance between a pair of points is

$$r = \sqrt{X^2 + Y^2} \ .$$

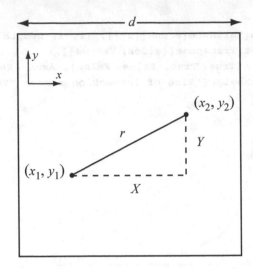

Figure 3.2. Coordinate system for determining probability density of separation of pairs of points

We seek an expression for the probability density of r, so must consider the distance between all possible pairs (x_i, y_i). Accordingly, we must apply Pythagoras' theorem stochastically and referring to Figure 3.2 it is clear that to do this we must obtain first the probability densities of the sides X and Y. We consider the points (x_i, y_i) occurring within a square region of side d where the random variables x_i and y_i have a uniform distribution on the interval $[0, d]$. The probability densities of x and y are the same and given by

$$f(x) = \frac{1}{d} \qquad \text{for } 0 \le x \le d \tag{3.1}$$

$$g(y) = \frac{1}{d} \qquad \text{for } 0 \le y \le d . \tag{3.2}$$

For a pair of points (x_1, y_1) and (x_2, y_2), the random variables X and Y are given by

$$X = |x_1 - x_2| \tag{3.3}$$
$$Y = |y_1 - y_2| . \tag{3.4}$$

Now, if the random variables j and k are independent with probability densities $f(j)$ and $g(k)$ in the domain $-\infty \le j, k \le \infty$, then the probability density of the random variable $l = j + k$ is given by

$$h(l) = \int_{-\infty}^{\infty} f(j)\, g(l - j)\, \mathrm{d}k . \tag{3.5}$$

Our random variables x and y are defined on the interval $[0, d]$ and we require X and Y to be positive, so we must apply Equation 3.5 in two parts with appropriate limits of integration. We have two cases,

$$X = \begin{cases} x_1 - x_2 & \text{when } x_2 \leq x_1 \\ x_2 - x_1 & \text{when } x_2 > x_1 \end{cases} . \tag{3.6}$$

So the probability density of X is given by

$$p(X) = \int_0^{d-X} f(x)\, f(X-x)\, \mathrm{d}x + \int_X^d f(x)\, f(X-x)\, \mathrm{d}x . \tag{3.7}$$

As x has a uniform probability density, $f(x)$ and $f(X-x)$ are both given by $1/d$, so the probability density of X is computed in *Mathematica* using the following code, where the last line is included as check that the probability density function we have derived integrates to 1 over the domain of X

```
In[1]:= pdfx = 1 / d;
       pdfX = Integrate[pdfx^2 , {x1, X, d}] +
          Integrate[pdfx^2, {x1, 0, d - X}]
       Integrate[pdfX, {X, 0, d}]
```

$$\text{Out[2]=}\quad \frac{2\,(d - X)}{d^2}$$

$$\text{Out[3]=}\quad 1$$

The probability density of Y is, of course,

```
In[4]:= pdfY = pdfX /. X → Y
```

$$\text{Out[4]=}\quad \frac{2\,(d - Y)}{d^2}$$

We seek the probability density of $r = \sqrt{X^2 + Y^2}$. We obtain this in three stages: first we derive the probability densities of X^2 and Y^2, then the probability density of their sum, and finally that of the square root of the sum. We define a new variable $X_{sq} = X^2$, which has probability density given by

$$q(X_{sq}) = p(X = \sqrt{X_{sq}}) \left| \frac{dX}{dX_{sq}} \right| , \tag{3.8}$$

where the second term is the Jacobian, which we first encountered on page 43.

```
In[5]:= pdfXsquared = D[√Xsq , Xsq] pdfX /. X → √Xsq
       PowerExpand[Integrate[pdfXsquared, {Xsq, 0, d^2}]]
```

Out[5]= $\dfrac{d - \sqrt{Xsq}}{d^2 \sqrt{Xsq}}$

Out[6]= 1

The probability density of $Y_{sq} = Y^2$ is

In[7]:= **pdfYsquared = pdfXsquared /. Xsq → Ysq**

Out[7]= $\dfrac{d - \sqrt{Ysq}}{d^2 \sqrt{Ysq}}$

To compute the probability density of $r^2 = X^2 + Y^2$, we must again apply Equation 3.5 choosing appropriate limits for the integrals. Intuitively, integrating over the range $0 \le X^2 \le d^2$ may seem appropriate, but specifying the problem in this way does not take account of the fact that for a given value of X^2, that of r^2 must exceed it. We overcome this by handling the integral piecewise for the intervals $0 \le r^2 < d^2$ and $d^2 \le r^2 < 2\,d^2$. Bearing in mind that $r^2 \ge X^2$ we derive the probability density of $r_{sq} = r^2$ for the first interval as follows:

In[8]:= **Integrate[pdfXsquared**
       (pdfYsquared /. Ysq → (Rsq - Xsq)), {Xsq, 0, Rsq}]

Out[8]= $\text{If}\bigg[Rsq > 0, \dfrac{d^2\,\pi - 4\,d\,\sqrt{Rsq} + Rsq}{d^4}, \text{Integrate}\bigg[$

$\dfrac{1}{d^4} - \dfrac{1}{d^3\,\sqrt{Rsq - Xsq}} - \dfrac{1}{d^3\,\sqrt{Xsq}} + \dfrac{1}{d^2\,\sqrt{Rsq - Xsq}\,\sqrt{Xsq}},$

$\{Xsq, 0, Rsq\}, \text{Assumptions} \to Rsq \le 0\bigg]\bigg]$

Mathematica provides an output in the form of an **If** statement, where the first term states the condition for which the second term is true, and the third term gives the integral unevaluated as the best output it can achieve if the condition is false. In this case, the condition is suitable for our system, so we re-evaluate the integral specifying the condition as an assumption:

In[9]:= **pdfRsquaredInterval1 = Integrate[**
       pdfXsquared (pdfYsquared /. Ysq → (Rsq - Xsq)),
       {Xsq, 0, Rsq}, Assumptions → Re[Rsq] > 0]

Out[9]= $\dfrac{d^2\,\pi - 4\,d\,\sqrt{Rsq} + Rsq}{d^4}$

For the second interval, we have:

In[10]:= **pdfRsquaredInterval2 = Integrate[**
    pdfXsquared (pdfYsquared /. Ysq → (Rsq - Xsq)),
    {Xsq, Rsq - d², d²}, Assumptions → 0 < d² < Rsq < 2 d²];
    pdfRsquaredInterval2 = FullSimplify[
    pdfRsquaredInterval2, d > 0]

Out[11]= $-\dfrac{1}{d^4}\left(Rsq - 4\,d\,\sqrt{-d^2 + Rsq} \;+\right.$

$$\left. 2\,d^2\left(1 + ArcCos\left[d\,\sqrt{\dfrac{1}{Rsq}}\right] - ArcSin\left[\dfrac{d}{\sqrt{Rsq}}\right]\right)\right)$$

In evaluating the last integral, we have stated the required assumptions. Often it is worthwhile to first evaluate the integral without stating assumptions and, when the output is in the form of an **If** statement, inspecting this before specifying appropriate assumptions. An alternative is to tell *Mathematica* to assume what is required to evaluate the integral by including the option **GenerateConditions->False**; this should be used with caution.

Having defined the probability density of $R_{sq} = r^2$ over two intervals, we bring these two expressions together using **Piecewise**, specifying the appropriate conditions for which each expression applies:

In[12]:= **pdfRsquared =**
    Piecewise[{{pdfRsquaredInterval1, 0 ≤ Rsq < d²},
    {pdfRsquaredInterval2, d² ≤ Rsq ≤ 2 d²}}]

Out[12]= $\begin{cases} \dfrac{d^2\,\pi - 4\,d\,\sqrt{Rsq} + Rsq}{d^4} & 0 \le Rsq < d^2 \\[2em] -\dfrac{Rsq - 4\,d\,\sqrt{-d^2 + Rsq} + 2\,d^2\left(1 + ArcCos\left[d\,\sqrt{\tfrac{1}{Rsq}}\right] - ArcSin\left[\tfrac{d}{\sqrt{Rsq}}\right]\right)}{d^4} & d^2 \le Rsq \le 2\,d^2 \end{cases}$

We integrate over the appropriate ranges and sum the outputs to check that the probability density integrates to 1:

```
In[13]:= Integrate[pdfRsquared, {Rsq, 0, d²},
           Assumptions → 0 < Rsq < d² && d > 0]
         Integrate[pdfRsquared, {Rsq, d², 2 d²},
           Assumptions → d² < Rsq < 2 d² && d > 0]
         Simplify[% + %%]
```

Out[13]= $\dfrac{1}{6} (-13 + 6\pi)$

Out[14]= $\dfrac{1}{6} (19 - 6\pi)$

Out[15]= 1

The mean value, $\overline{r^2}$ is computed in a similar way:

```
In[16]:= Integrate[Rsq pdfRsquared, {Rsq, 0, d²},
           Assumptions → 0 < Rsq < d² && d > 0]
         Integrate[Rsq pdfRsquared, {Rsq, d², 2 d²},
           Assumptions → d² < Rsq < 2 d² && d > 0]
         Rsqbar = Simplify[% + %%]
```

Out[16]= $\dfrac{1}{30} d^2 (-38 + 15\pi)$

Out[17]= $-\dfrac{1}{10} d^2 (-16 + 5\pi)$

Out[18]= $\dfrac{d^2}{3}$

We seek the probability density of $r = \sqrt{R_{sq}}$ so must transform variables using

$$f(r) = g(R_{sq} = r^2) \left| \frac{dRsq}{dr} \right| . \qquad (3.9)$$

```
In[19]:= Rsq = r²;
         pdfR = D[Rsq, r] pdfRsquared;
         FullSimplify[PowerExpand[PiecewiseExpand[pdfR]]]
         Rsq =.
```

$$\text{Out[21]=} \quad \begin{cases} \dfrac{2\,r\,\left(d^2\,\pi - 4\,d\,r + r^2\right)}{d^4} & d^2 > r^2 \\[2em] -\dfrac{2\,r\,\left(r^2 - 4\,d\,\sqrt{-d^2 + r^2} + d^2\,\left(2 + \pi - 4\,\text{ArcSin}\left[\frac{d}{r}\right]\right)\right)}{d^4} & 2\,d^2 \ge r^2 \,\&\&\, d^2 \le r^2 \end{cases}$$

Again, we check that the probability density we have derived integrates to 1 over the domain:

```
In[23]:= Simplify[
           Integrate[pdfR, {r, 0, d}, Assumptions → 0 < r < d] +
             Integrate[pdfR, {r, d, √2 d},
               Assumptions → d <= r < √2 d]]
```

Out[23]= 1

The expected separation, $\bar{r}$, between pairs of points is given by:

```
In[24]:= rbar = FullSimplify[
           Integrate[r pdfR, {r, 0, d}, Assumptions → 0 < r < d] +
             Integrate[r pdfR, {r, d, √2 d},
               Assumptions → d <= r < √2 d]]
```

$$\text{Out[24]=} \quad \frac{1}{15}\,d\,\left(2 + \sqrt{2} + 5\,\text{ArcSinh}[1]\right)$$

and the variance $\sigma^2(r) = \overline{r^2} - \bar{r}^2$, so we have,

```
In[25]:= rvar = Simplify[Rsqbar - rbar²]
```

$$\text{Out[25]=} \quad \frac{1}{225}\,d^2\,\left(75 - \left(2 + \sqrt{2} + 5\,\text{ArcSinh}[1]\right)^2\right)$$

As a check, we compute r for a million pairs of random numbers occurring in a unit square.

```
In[26]:= d = 1;
         n = 1 000 000;
         SeedRandom[1]
         X = Abs[RandomReal[{0, d}, n] - RandomReal[{0, d}, n]];
         Y = Abs[RandomReal[{0, d}, n] - RandomReal[{0, d}, n]];
         R = √(X² + Y²) ;
```

We observe good agreement between our numerical estimate of the mean and variance and that we have determined analytically.

```
In[32]:= {{Mean[R], N[rbar]}, {Variance[R], N[rvar]}}
         Needs["Histograms`"]
         Show[Histogram[R, HistogramCategories → 30,
           HistogramScale → 1], Plot[pdfR,
           {r, 0, √2 d}, PlotStyle → Thickness[.005]],
           AxesLabel → {"r", "f(r)"}]
         d =.
```

Out[32]= {{0.521519, 0.521405}, {0.0614931, 0.0614697}}

Out[34]=
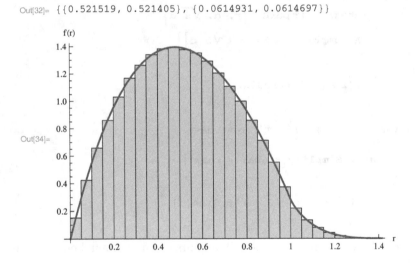

We state then that the probability density of the distance between pairs
of points chosen independently and at random in a square of side d is

$$
b(r,d) = \begin{cases}
\frac{2r}{d^4}\left(\pi d^2 - 4dr + r^2\right) & \text{for } 0 \le r \le d \\
\begin{aligned}
\frac{2r}{d^4}\big(&4d\sqrt{r^2 - d^2} - r^2 \\
&-d^2\left(2 + \pi - 4\sin^{-1}(\tfrac{d}{r})\right)\big)
\end{aligned} & \text{for } d \le r \le \sqrt{2}d \quad (3.10) \\
0 & \text{otherwise.}
\end{cases}
$$

In Section 4.2.3, we will use this probability density function to model
the occurrence of fibre centres within square inspection zones that partition
the plane of a fibre network. Dacey [26] notes that the probability density
we have obtained was derived independently by Robbins [134], Garwood [55],
and Ghosh [57]. These derivations required a transformation from Cartesian to
polar coordinates to handle the integrals involved. An alternative derivation,
which provides the probability density for the separation of pairs of points in
a rectangle of sides a and b, corresponding to the case considered by Ghosh,
was provided more recently without transformation to polar coordinates by
Bettstetter *et al.* [8]. Interestingly, Bettstetter *et al.* were interested in the
distribution of distances between points in the context of a rather different
stochastic network from those which occupy us: wireless communication networks.

3.3 Poisson Line Processes

The clustering of fibre centres influences the extent to which fibres interact
with each other within our network and the manner in which this interaction
differs between regions. Our standard model for a fibre in a two-dimensional
network is a rectangle of length λ and width ω and these may have a distribution of lengths and widths. Before considering such objects as the structural
element of our network, it is illustrative to consider first networks of *lines*,
reducing our fibres to one-dimensional elements with width zero. It is convenient first to consider lines of infinite length, so our random network consists of infinite lines with uniformly distributed orientation and each passing
through a point distributed according to a Poisson point process in a plane.
Many properties of such random line networks are given in the seminal work
of Miles [102]. We shall encounter and use several of Miles' results in what
follows, but first we note some intrinsic properties of the network that allow
us to avoid unnecessary assumptions.

- No pair of lines has the same orientation, so all pairs of lines will
  intersect.
- No group of three lines intersects at a common point.
- The points of intersection between lines constitute a point Poisson
  process in the plane.

We begin our treatment by generating a graphical representation of a random process of 50 lines occurring in a unit square to guide our subsequent analysis. The equation of a line with orientation θ and passing through a point (x_1, y_1) is

$$y = \tan(\theta)\,(x - x_1) + y_1 \;, \tag{3.11}$$

so, to generate random equations of lines for plotting we must generate random θ_i, x_i and y_i, each of which is a uniformly distributed continuous random variable.

```
In[1]:= n = 50;
        SeedRandom[1]
        θi = RandomReal[{0, π}, n];
        xi = RandomReal[{0, 1}, n];
        yi = RandomReal[{0, 1}, n];
```

Mathematica handles lists very efficiently, and operates in parallel on list elements, so to apply Equation 3.11 we need only to input

```
In[6]:= lines = Tan[θi] (x - xi) + yi;
```

and *Mathematica* will automatically use the ith element of each component list to generate the ith element of the resultant list. Naturally, list operations of this type can only be performed on lists of the same dimensions. We generate our graphic using **Plot**:

```
In[7]:= Plot[lines, {x, 0, 1},
        PlotRange → {{0, 1}, {0, 1}}, AspectRatio → Automatic,
        Frame → True, PlotStyle → Black]
```

Out[7]=

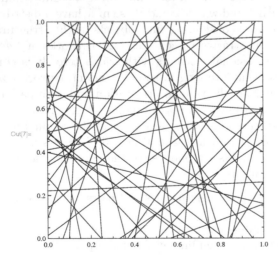

We observe that the line process has partitioned our square region into polygons each of which is defined by the line segments occurring between crossings. In materials such as non-woven textiles, the fibre staple is sufficiently long that for many purposes infinite lines provide quite a good approximation of the network structure. In the case of electrospun networks, the fibre is produced as a continuous filament, and we might expect good agreement for these materials also. It is also immediately apparent that network features such as crossings, line segment lengths and polygon size are closely coupled; regions with many crossings exhibiting shorter segment lengths and smaller polygons and *vice versa*. In the next sections we consider some of these dependencies in more detail.

3.3.1 Process Intensity

Each of the lines in the graphic we have generated has different length and so has a different contribution to the structure of the network; longer lines generating more crossings within the unit square and thus being associated with more polygons. Accordingly, the number of lines occurring within a square is not a good descriptor of the intensity of our line process. A better measure is the expected total length of lines per unit area, and we derive an expression for this here.

Consider a line with orientation θ occurring within a unit square, as illustrated in Figure 3.3. The distribution of orientations of lines θ is uniform between 0 and π so we can make use of symmetry and consider only $0 \le \theta \le \frac{\pi}{2}$. We observe from Figure 3.3 that lines A and C intersect adjacent sides of the square and line B intersects opposite sides; no other configurations are possible.

If the coordinates of the points of intersection with the perimeter of the square are uniformly distributed random variables $0 \le x, y \le 1$ then we may state that two thirds of lines occurring independently and at random within the unit square have length,

$$l_1 = \sqrt{x_i^2 + y_i^2}\,,$$

and one third have length,

$$l_2 = \sqrt{(x_i - x_j)^2 + 1}\,.$$

We proceed then to derive the probability densities of l_1 and l_2 in order that we may combine these to give the probability density of the length of lines occurring within the unit square and hence the mean. As might be expected, the derivation is rather similar to that we gave earlier for the probability density of the separation of pairs of points.

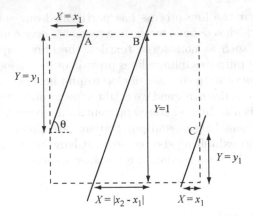

Figure 3.3. Schema for possible configurations of a line with orientation θ intersecting the sides of a square region

Case 1: Lines of types A and C

The probability densities of x and y, $f(x) = f(y) = 1$ so the probability densities of $X_{sq} = x^2$ and $Y_{sq} = y^2$ are

In[1]:= **pdfXsquared = D$\left[\sqrt{\textbf{Xsq}}, \textbf{Xsq}\right]$**

pdfYsquared = pdfXsquared /. Xsq → Ysq

Out[1]= $\dfrac{1}{2\sqrt{\textbf{Xsq}}}$

Out[2]= $\dfrac{1}{2\sqrt{\textbf{Ysq}}}$

As previously, we compute the probability density of $l_1^2 = x^2 + y^2$ over two intervals, combine these into a piecewise function and check that this integrates to 1 over the domain.

```
In[3]:= pdfLsquaredInterval1 =
        Integrate[pdfXsquared (pdfYsquared /. Ysq → (Lsq - Xsq)),
        {Xsq, 0, Lsq}, Assumptions → Re[Lsq] > 0];
        pdfLsquaredInterval2 = Integrate[
           pdfXsquared (pdfYsquared /. Ysq → (Lsq - Xsq)),
           {Xsq, Lsq - 1, 1}, Assumptions → 1 < Lsq < 2 ];
        pdfLsquaredCase1 = Piecewise[{{pdfLsquaredInterval1,
           0 ≤ Lsq < 1}, {pdfLsquaredInterval2, 1 ≤ Lsq ≤ 2}}]
        Simplify[Integrate[pdfLsquaredCase1, {Lsq, 0, 1},
           Assumptions → 0 < Lsq < 1] + Integrate[pdfLsquaredCase1,
           {Lsq, 1, 2}, Assumptions → 1 < Lsq < 2]]
```

$$
\text{Out[5]=}\quad
\begin{cases}
\dfrac{\pi}{4} & 0 \le \text{Lsq} < 1 \\[2mm]
\dfrac{1}{2}\left(\text{ArcCsc}\left[\sqrt{\text{Lsq}}\,\right] - \text{ArcTan}\left[\sqrt{-1+\text{Lsq}}\,\right]\right) & 1 \le \text{Lsq} \le 2
\end{cases}
$$

Out[6]= 1

Case 2: Lines of type B

Since $f(x) = 1$, the probability density of $X = |x_i - x_j|$ is

```
In[7]:= pdfX =
        Integrate[1, {x, X, 1}] + Integrate[1, {x, 0, 1 - X}]
```

Out[7]= 2 - 2 X

and the probability density of $X_{sq} = X^2$ is

```
In[8]:= pdfXsquared = Simplify[D[√Xsq , Xsq] pdfX /. X → √Xsq ]
```

$$\text{Out[8]=}\quad -1 + \frac{1}{\sqrt{\text{Xsq}}}$$

To obtain the probability density of l_2^2 we need to perform the variable transform for $X^2 = l^2 - 1$ and check that this integrates to 1.

```
In[9]:= pdfLsquaredCase2 =
        Piecewise[{{pdfXsquared /. Xsq → (Lsq - 1), 1 ≤ Lsq ≤ 2}}]
        Integrate[pdfLsquaredCase2, {Lsq, 1, 2},
           Assumptions → 1 < Lsq < 2]
```

$$\text{Out[9]}= \begin{cases} -1 + \dfrac{1}{\sqrt{-1+\text{Lsq}}} & 1 \le \text{Lsq} \le 2 \end{cases}$$

$$\text{Out[10]}= 1$$

Combined distribution

In combining the distributions, we must take account of the fact that lines of types A and C occur with twice the frequency as those of type B. The probability density of $L_{sq} = l^2$ for all lines is therefore given by:

```
In[11]:= pdfLsquared = FullSimplify[PiecewiseExpand[
              (2 / 3) pdfLsquaredCase1 + (1 / 3) pdfLsquaredCase2]]
```

$$\text{Out[11]}= \begin{cases} \dfrac{\pi}{6} & 0 \le \text{Lsq} < 1 \\[2mm] \dfrac{1}{3}\left(-1 + \dfrac{1}{\sqrt{-1+\text{Lsq}}} + \text{ArcCsc}\left[\sqrt{\text{Lsq}}\,\right] - \text{ArcTan}\left[\sqrt{-1+\text{Lsq}}\,\right]\right) & 1 \le \text{Lsq} \le 2 \end{cases}$$

We obtain the probability density of $l = \sqrt{L_{sq}}$ in the usual way.

```
In[12]:= Lsq = 1²;
         pdfL = D[Lsq, 1] pdfLsquared;
         pdfL =
         FullSimplify[PowerExpand[PiecewiseExpand[pdfL]], 1 ≥ 0]
```

$$\text{Out[14]}= \begin{cases} \dfrac{1\,\pi}{3} & 1 < 1 \\[2mm] \dfrac{2}{3}\,1\left(-1 + \dfrac{1}{\sqrt{-1+1^2}} + \text{ArcCsc}[1] - \text{ArcTan}\left[\sqrt{-1+1^2}\,\right]\right) & 1 \le 1 \le \sqrt{2} \end{cases}$$

A plot of the probability density of r reveals a discontinuity at $l = 1$.

```
In[15]:= Plot[pdfL, {1, 0, √2}, AxesLabel → {"1", "f(1)"}]
```

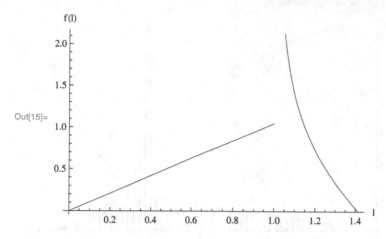

Out[15]=

Given such an unusual probability density, it is wise to carry out a Monte Carlo simulation and generate a histogram of the frequency of line lengths against which we can compare our expression.

```
In[16]:= n = 1 000 000;
        SeedRandom[1]
        X1 = RandomReal[{0, 1}, n];
        Y1 = RandomReal[{0, 1}, n];
        L1 = √(X1² + Y1²) ;
        X2 =
          Abs[RandomReal[{0, 1}, n / 2] - RandomReal[{0, 1}, n / 2]];
        L2 = √(X2² + 1) ;
        L = Join[L1, L2];
        Needs["Histograms`"]
        Show[
          Histogram[L, HistogramCategories → 50, HistogramScale → 1],
          Plot[pdfL, {1, 0, √2 }, PlotStyle → Thickness[.005],
            PlotRange → {All, {0, 6}}], AxesLabel → {"1", "f(1)"}]
```

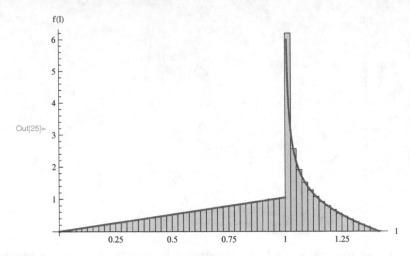

The agreement between the probability density we have derived and that obtained numerically is excellent so we may proceed with confidence to calculate the expected length of a line occurring randomly within a unit square.

In[26]:= `lbar = Simplify[`
    `Integrate[1 pdfL, {1, 0, 1}, Assumptions → 0 ≤ 1 ≤ 1] +`
        `Integrate[1 pdfL, {1, 1, √2},`
        `Assumptions → 1 ≤ 1 ≤ √2]]`

Out[26]= $\dfrac{1}{18}\left(4 + 2\sqrt{2} + 4\,\mathrm{ArcSinh}[1] + 3\,\mathrm{Log}\left[3 + 2\sqrt{2}\right]\right)$

The expression we have obtained for $\bar{l}$ includes an inverse hyperbolic sine term and a logarithm, suggesting that a simpler form can be obtained. Applying **FullSimplify** however, does not greatly simplify the output.

In[27]:= `FullSimplify[lbar]`

Out[27]= $\dfrac{1}{18}\left(4 + 2\sqrt{2} + 4\,\mathrm{ArcSinh}[1] + \mathrm{Log}\left[99 + 70\sqrt{2}\right]\right)$

A simpler expression can be obtained by converting the inverse hyperbolic sine to its logarithmic form using **TrigToExp**:

In[28]:= **TrigToExp[lbar]**

Out[28]= $\frac{2}{9} + \frac{\sqrt{2}}{9} + \frac{2}{9} \text{Log}\left[1 + \sqrt{2}\right] + \frac{1}{6} \text{Log}\left[3 + 2\sqrt{2}\right]$

Inspection of the resultant expression reveals that **FullSimplify** would not generate a simpler form because $\sqrt{2}$ can take negative values yielding complex solutions. Observing also that $(3 + 2\sqrt{2}) = (1 + \sqrt{2})^2$, we make the substitution $z = (1 + \sqrt{2})$ and simplify this, telling *Mathematica* that $z > 0$; substituting again for z and simplifying yields our final simplified expression for $\bar{l}$.

In[29]:= **lbar = FullSimplify$\left[$Simplify$\left[\right.$**

$$\frac{2}{9} + \frac{\sqrt{2}}{9} + \frac{2\,\text{Log}[z]}{9} + \frac{\text{Log}\left[z^2\right]}{6}, \; z > 0\right] /. \, z \rightarrow \left(1 + \sqrt{2}\right)\Big]$$

Out[29]= $\frac{1}{9}\left(2 + \sqrt{2} + 5\,\text{ArcSinh}[1]\right)$

For completeness, we compute the variance of line lengths within our unit square, and note that our numerical estimates agree quite closely with that we have obtained analytically.

In[30]:= **lvar = Simplify$\left[$Integrate$\left[l^2\,\text{pdfL}, \; \{1, \, 0, \, 1\},\right.\right.$**

  Assumptions $\rightarrow$ 0 $\le$ 1 $\le$ 1$\Big]$ + Integrate$\Big[l^2\,$pdfL,

  $\left\{1, \, 1, \, \sqrt{2}\,\right\}$, **Assumptions $\rightarrow$ 1 $\le$ 1 $\le$ $\sqrt{2}\,\Big]\Big]$ - lbar^2**

Out[30]= $\frac{5}{6} - \frac{1}{81}\left(2 + \sqrt{2} + 5\,\text{ArcSinh}[1]\right)^2$

In[31]:= **{{N[lbar], Mean[L]}, {N[lvar], Variance[L]}}**

Out[31]= {{0.869009, 0.868962}, {0.0781566, 0.0780233}}

We may now describe the intensity of our process in terms of the expected total line length per unit area, τ. For a process of n lines per unit area is given by

$$\tau = n\bar{l}\,. \tag{3.12}$$

We will derive the expected number of crossings per unit area for networks of fibres with finite length in Section 4.2.1. The resultant expression is the same as that provided by Miles [102] for Poisson line processes:

$$n_{\text{cross}} = \frac{\tau^2}{\pi} . \tag{3.13}$$

So, the number of crossings per unit area depends on the process intensity only.

Example Calculation

Consider a glass fibre network of the type often used in composites and formed using a non-woven process. If such a glass fibre veil has mean mass per unit area, $\bar{\beta}$ and it is formed from fibres with linear density δ, then the expected length of fibres per unit area is,

$$\tau = \frac{\bar{\beta}}{\delta} . \tag{3.14}$$

For fibres of density ρ and circular cross section with diameter ω, the linear density is given by,

$$\delta = \frac{\pi \omega^2}{4} \rho \tag{3.15}$$

The *Mathematica* package **"Units`"** conveniently handles the units of variables for us and allows conversion between these. For a glass fibre veil with mean mass per unit area 10 $\text{g}\,\text{m}^{-2}$ formed from E-glass fibres with width 10 μm we first calculate the linear density of fibres using Equation 3.15.

```
In[1]:= Needs["Units`"]
        β = 10 Gram/Meter²; ω = 10 Micro Meter;
        ρ = 2550 Kilogram/Meter³;
        δ = SI[π ω² ρ / 4.]
```

$$Out[3]= \frac{2.00277 \times 10^{-7} \text{ Kilogram}}{\text{Meter}}$$

We compute the process intensity using Equation 3.14:

```
In[4]:= τ = SI[β / δ]
```

$$Out[4]= \frac{49\,931.}{\text{Meter}}$$

The resultant number is rather striking: our network contains about 50 km of fibre per square metre. The expected number of crossings per unit area is given by Equation 3.13:

In[5]:= **ncross = τ^2 / π**

Convert$[$ncross, (Centimeter)$^{-2}]$

Out[5]= $\dfrac{7.93579 \times 10^8}{\text{Meter}^2}$

Out[6]= $\dfrac{79\,357.9}{\text{Centimeter}^2}$

So our network has about 8×10^4 fibre intersections per square centimetre and this represents the number of sites at which binding agents may act, for example, to bond fibres to each other to provide network strength.

3.3.2 Inter-crossing Distances

Our random lines partition the plane into polygons and, on any given line, the distances between its points of intersection with other lines define the sides of these polygons. The manufacture and processing of planar fibre networks typically involve the web being placed under strain in the direction of processing and these inter-crossing ligaments form the load-bearing elements of the structure.

The mean inter-crossing distance or ligament length is given by the expected line length per unit area divided by the expected number of crossings per unit area, *i.e.*

$$\bar{g} = \frac{\tau}{n_{\text{cross}}} \, . \tag{3.16}$$

So from Equation 3.13 we have,

$$\bar{g} = \frac{\pi}{\tau} \, . \tag{3.17}$$

So the expected distance between crossings is inversely proportional to the intensity of the line process.

Inevitably, we have a distribution of ligament lengths in our network and we derive the probability density of these by considering the incidence of crossings along an arbitrary line. We have seen that the occurrence of crossings within a plane represents a point Poisson process in two dimensions. Here we consider the corresponding one-dimensional process, where the events are the occurrence of crossings along a line and the expected number of crossings per unit length is μ. The expected number of crossings in an interval of length g is μg and the probability of there being no intersections in an interval of length g is the Poisson probability of zero events in our process, *i.e.*

In[1]:= **Pzero = PDF$[$PoissonDistribution$[\mu\, g]$, 0$]$**

Out[1]= $e^{-g\,\mu}$

We require the probability density of g which must integrate to unity over the domain $0 \leq 0 < \infty$. Accordingly, the probability density of g is given by

In[2]:= **pdfg =**
      Pzero / Integrate[Pzero, {g, 0, ∞}, Assumptions → μ > 0]

Out[2]= $e^{-g\,\mu}\,\mu$

The mean $\bar{g}$ is given by

In[3]:= **Integrate[g pdfg, {g, 0, ∞}, Assumptions → μ > 0]**

Out[3]= $\dfrac{1}{\mu}$

Substituting for μ, we obtain the probability density for g in terms of the mean of the distribution only:

In[4]:= **pdfg /. μ → (1 / gbar)**

Out[4]= $\dfrac{e^{-\frac{g}{gbar}}}{gbar}$

We recognise this as the probability density function for the exponential distribution as introduced on page 45.

In[5]:= **PDF[ExponentialDistribution[1 / gbar], g]**
      Mean[ExponentialDistribution[1 / gbar]]
      Variance[ExponentialDistribution[1 / gbar]]

Out[5]= $\dfrac{e^{-\frac{g}{gbar}}}{gbar}$

Out[6]= gbar

Out[7]= $gbar^2$

We state then, that the distribution of ligament lengths in a random process of infinite lines is exponential with mean $\bar{g}$ and variance $\sigma^2(g) = \bar{g}^2$.

Though the derivation is simple, the result is important. Firstly, we have shown that a continuous probability density can be obtained from a discrete probability distribution. Secondly, though we have derived the exponential distribution here in the context of the ligament length distribution, it has broad applicability to the waiting times between events in arbitrary random processes.

Papoulis [119] notes that if the intervals between consecutive points of a point process are independent and exponentially distributed, then this process is Poisson. This is a valuable observation since it provides a mechanism to determine how close a given network is to having a random structure. Widely available image analysis software permits rapid measurement of the distances between fibre crossings on arbitrary lines drawn on a micrograph of a planar fibre network. If the distribution of the gaps between fibres is close to exponential, then the network has a structure close to that realised by a random process. Analysis of this type for thin networks of papermaking fibres is reported by Kallmes and Corte [74]; they report exponential gap length distributions despite the fibres having width and finite length, neither of which we have considered here. We will address problems of finite fibres in due course, but for now we note the comment of Deng and Dodson [27] that for random networks of rectangles with length λ and width ω the exponential distribution provides a good approximation to the distribution of ligament lengths provided that $\lambda \gg \omega$ and $\lambda/\bar{g} \gg 1$, which would have been the case for the networks studied by Kallmes and Corte.

3.3.3 Statistics of Polygons

The polygons arising from the partitioning of the plane by random lines have relevance to pore size distribution of planar fibrous materials and hence are important for the mechanical entrapment of particles by fibrous filters [1, 11] and the ingress of cells into electrospun scaffolds used in tissue engineering [92, 121].

When we examine a graphical representation of a random line process such as that we generated on page 72 we observe that the polygons generated are all convex, *i.e.* lines connecting any pair of vertices are contained within the polygon. This means of course that the internal angles of polygons exist in the interval $0 < \phi < \pi$. Miles [102] gives the probability density of intersection angles as

$$f(\phi) = \frac{1}{2} \sin(\phi) \qquad (3.18)$$

We state this without derivation here, though we will derive the probability density of the angles of intersection between lines of finite length in Section 4.2.1. Both Miles [102] and George [56] note that the form of Equation 3.18 is somewhat counterintuitive, since each pair of lines forms an intersection and generates a pair of intersection angles ϕ and $(\pi - \phi)$. The reducing probability density as ϕ approaches 0 and π arises because the points of intersection 'are shifted towards infinity' [56].

Bearing in mind that no set of three lines intersects at a common point, we observe from Figure 3.4 that each point of intersection generates a vertex of four polygons. It follows that the expected number of sides per polygon is

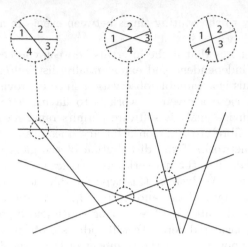

Figure 3.4. Each crossing between pairs of lines is associated with four polygons

four also. For completeness, we state the result of Miles [102] that the variance of the number of sides per polygon is $\left(\pi^2 - 8\right)/2$.

A natural extension of the expected number of sides per polygon and the number of polygons having vertices at each intersection being four, is that the expected number of polygons per unit area is the same as the expected number of intersections between lines, *i.e.*

$$n_{\mathrm{poly}} = n_{\mathrm{cross}} \tag{3.19}$$

So the expected area of a polygon is

$$\bar{a}_{\mathrm{poly}} = \frac{1}{n_{\mathrm{cross}}}$$
$$= \frac{\pi}{\tau^2} \cdot \tag{3.20}$$

Polygon Side Distribution

If we consider an evolving line process, then each additional line partitions the polygons arising from those that have been placed previously. Figure 3.5 shows some irregular convex n-sided polygons along with the polygons arising from their partitioning. If the partitioning line intersects adjacent sides of the original n-gon, then the resulting polygons are a triangle and a polygon of $(n+1)$ sides; otherwise, the partition yields polygons with between four and n sides. Accordingly, we expect the population of polygons to be dominated by triangles and quadrilaterals, and it turns out that this is the case. The precise distribution of polygon sides is an outstanding theoretical problem, however, Miles [102] gives the probability of triangles:

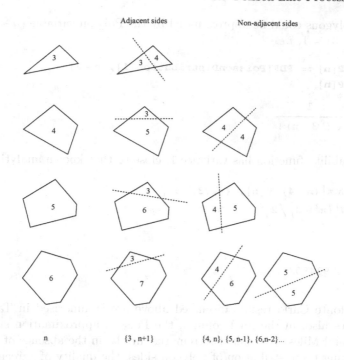

Figure 3.5. Polygons arising from the partitioning of n-gons by lines passing through adjacent and non-adjacent sides

$$P(n = 3) = 2 - \frac{\pi^2}{6} \approx 0.355 \; ,$$

and the probability of quadrilaterals was derived almost 20 years later by Tanner [159]:

$$P(n = 4) = -\frac{1}{3} - \frac{7\pi^2}{36} + 4 \int_0^{\pi/2} x^2 \cot(x) \, \mathrm{d}x \approx 0.381 \; .$$

So almost 74 % of the polygons are either triangles or quadrilaterals and the latter are more common than other polygons.

The fractions of polygons with more than four sides have been determined using Monte Carlo methods by several groups. Piekaar and Clarenburg [124] counted polygons arising from random processes of lines with *finite* length and proceeded to develop theory describing flow through aerosol filters [16, 17]. Crain and Miles [22] extracted polygons arising from Poisson line processes and computed properties for 2×10^5 polygons. Later George [56] developed an algorithm for generating individual polygons without first generating a random line process and, given the improved computational efficiency of such a routine, obtained statistics characterising 2.5×10^6 polygons and observed no polygons with $n > 12$. Crain and Miles observed also that the probability of

n-sided polygons could be approximated by the Poisson variable $(n-3)$ with mean $(\bar{n}-3) = 1$, *i.e.*

```
In[1]:=  P[n] := PDF[PoissonDistribution[1], n-3]
         P[n]
```

$$\text{Out[2]}= \frac{1}{e \ (-3+n) \ !}$$

This probability function has variance 1, close to that known analytically:

```
In[3]:=  Sum[(n-4)² P[n], {n, 3, ∞}]
         N[(π² - 8) / 2]
```

Out[3]= 1

Out[4]= 0.934802

The Monte Carlo results discussed above are summarised in Table 3.1, which gives also, in the final column, the Poisson approximation suggested by Crain and Miles. We make two remarks: firstly, in the absence of analytic results giving the distribution of polygon sides, the quality of agreement between the numerical data and the Poisson approximation is sufficiently good that we may use the Poisson approximation in subsequent treatments with confidence that any error is small. Secondly, we observe reasonable agreement between the data of Piekaar and Clarenburg for finite line processes and those for infinite line processes. Accordingly, to a reasonable degree of approximation, we may assume that the polygon size distribution is unaffected between these cases.

Table 3.1. Collection of Monte Carlo results for the probability that random polygons have n sides. Final column gives Poisson approximation of Crain and Miles [22]

	$P(n)$			
	Piekaar &	Crain &	George	$P(n-3)$
n	Clarenburg [124]	Miles [22]	[56]	
3	0.393	0.3558	0.3552	0.3679
4	0.369	0.3759	0.3814	0.3679
5	0.159	0.1889	0.1895	0.1839
6	0.070	0.0608	0.0587	0.0613
7	0.008	0.0130	0.0128	0.0153
8		0.0021	0.0021	0.0031
9		3.0×10^{-4}	2.7×10^{-4}	5.1×10^{-4}
10		2.5×10^{-5}	1.8×10^{-5}	7.2×10^{-5}
11		3.2×10^{-6}	2.8×10^{-6}	9.1×10^{-6}
12		2.4×10^{-6}	0.4×10^{-6}	1.0×10^{-6}

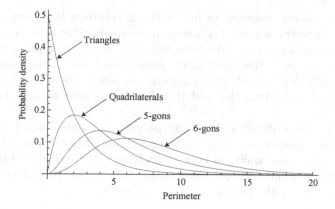

Figure 3.6. χ^2 probability density of perimeter lengths for random polygons

Perimeter and Area Distributions

The distribution of polygon perimeters is unknown analytically. For n-sided polygons however, Miles [102] shows that the perimeter has a χ^2-distribution with $2(n-2)$ degrees of freedom. We have not encountered this distribution before; it is a special case of the gamma distribution such that a χ^2-distribution with ν degrees of freedom is a gamma distribution with mean ν and coefficient of variation $\sqrt{2/\nu}$.

In[1]:= **PDF[ChiSquareDistribution[ν], p]**
      TrueQ[
        ChiSquareDistribution[ν] == GammaDistribution[ν / 2, 2]]

Out[1]= $$\frac{2^{-\nu/2} \, e^{-p/2} \, p^{-1+\frac{\nu}{2}}}{\text{Gamma}\left[\frac{\nu}{2}\right]}$$

Out[2]= **True**

It follows that the distribution of perimeters of triangles has a χ^2-distribution with $\nu = 2$ degrees of freedom, which is the same as a gamma distribution with coefficient of variation 1. Recall that the exponential distribution is a special case of the gamma distribution with coefficient of variation 1, so we may state that the distribution of perimeters of triangles is exponential.

The probability density of the perimeters of polygons, as given by the χ^2-distribution is shown in Figure 3.6 for polygons with sides $n \leq 6$. The distribution as we have stated it gives the perimeter in units of the mean inter-crossing distance. We note that the mean perimeter of an n-sided polygon is $2(n-2)$. Accordingly, the mean perimeter of quadrilaterals is four times the mean inter-crossing distance, in line with expectation; the mean perimeter of triangles however is twice the mean inter-crossing distance and not three times, as we might expect. We note also that for $n > 4$ the perimeter is greater than n times the mean inter-crossing distance. This, coupled with the probability densities shown in Figure 3.6, allows us to state that small polygons tend to be triangles and that larger polygons tend to have more sides [27, 102].

Although the probability density of polygon perimeters is not known analytically, an approximate probability density function has been derived recently [41]. If the probability of polygons with n sides is $P(n)$ and the probability density of perimeters of n-sided polygons is $q(p, n)$, then the probability density function for the perimeter of all polygons is

$$q(p) = \sum_{n=3}^{\infty} P(n) \, q(p, n) \, . \tag{3.21}$$

Since the probability function $P(n)$ is unknown, we may use the Poisson approximation of Crain and Miles [22]

In[3]:= **P[n] := PDF[PoissonDistribution[1], n - 3]**

The probability density $q(p, n)$ is χ^2 with $2(n - 2)$ degrees of freedom:

In[4]:= **v = 2 (n - 2);**
q[p_, n_] := PDF[ChiSquareDistribution[v], p]

We obtain the probability density $q(p)$ as given by Equation 3.21 using **Sum**, and check that the resultant expression integrates to 1:

In[6]:= **pdfp = Sum[P[n] q[p, n], {n, 3, ∞}]**
Integrate[pdfp, {p, 0, ∞}]

Out[6]= $\dfrac{1}{2} \, e^{-1-\frac{p}{2}} \, \text{BesselI}\!\left[0, \, \sqrt{2} \, \sqrt{p}\,\right]$

Out[7]= 1

The probability density we have obtained includes the special function **BesselI**, which is the modified Bessel function of the first kind. We obtain the mean and variance in the usual way:

In[8]:= **pbar = Integrate[p pdfp, {p, 0, ∞}]**
pvar = Integrate[(p - pbar)^2 pdfp, {p, 0, ∞}]

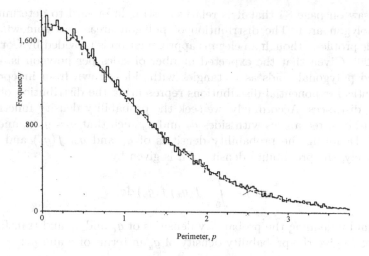

Figure 3.7. Monte Carlo data of Crain and Miles [22] for distribution of polygon perimeters. (I.K. Crain and R.E. Miles. *J. Statist. Comput. Simul.* 4:293-325, 1976. Copyright Taylor and Francis Ltd., www.informaworld.com. Reproduced with permission)

Out[8]= 4

Out[9]= 12

A plot of the probability density we have derived shows $q(p)$ to decay monotonically. Qualitatively, this compares favourably with the Monte Carlo data of Crain and Miles [22], as shown in Figure 3.7.

In[10]:= `Plot[pdfp, {p, 0, 20}, AxesLabel → {"p", "q(p)"}]`

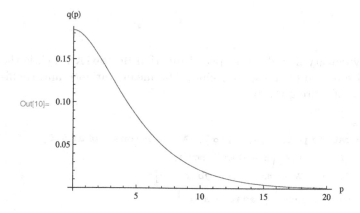

We saw on page 84 that it is relatively straightforward to determine the mean polygon area. The distribution of polygon areas is a somewhat less tractable problem, though an elegant approximation is provided by Corte and Lloyd [20]. Given that the expected number of sides per polygon is 4, they modelled polygonal voids as rectangles with sides drawn from independent and identical exponential distributions representing the distribution of inter-crossing distances. Accordingly, we seek the probability density function for the areas, a of rectangles with sides g_x and g_y such that $a = g_x g_y$ and $\bar{g}_x = \bar{g}_y = \bar{g}$. Denoting the probability densities of g_x and g_y, $f(g_x)$ and $f(g_y)$, respectively, the probability density of a is given by

$$g(a) = \int_0^\infty f(g_x) f(g_y) \, \mathrm{d}g_x \, . \tag{3.22}$$

We begin by defining the probability densities of g_x and g_y and transforming variables to give the probability density of g_y in terms of a and g_x:

```
In[1]:=  pdfgx = PDF[ExponentialDistribution[1 / gbar], gx];
         pdfgy = PDF[ExponentialDistribution[1 / gbar], gy];
         gy = a / gx;
         pdfgy = D[gy, a] pdfgy;
```

To obtain the probability density of a we evaluate Equation 3.22.

```
In[5]:=  pdfa = Integrate[pdfgx pdfgy,
              {gx, 0, ∞}, Assumptions → gbar > 0 && a ≥ 0]
         TrueQ[Integrate[pdfa, {a, 0, ∞},
              Assumptions → gbar > 0] == 1]
```

$$\text{Out[5]=} \quad \frac{2 \, \mathrm{BesselK}\left[0, \, \frac{2\sqrt{a}}{gbar}\right]}{gbar^2}$$

Out[6]= True

This probability density includes the special function **BesselK** which is the modified Bessel function of the second kind. The mean, variance and coefficient of variation of a are given by

```
In[7]:=  abar =
            Integrate[a pdfa, {a, 0, ∞}, Assumptions → gbar > 0]
         avar = Integrate[(a - abar)² pdfa,
              {a, 0, ∞}, Assumptions → gbar > 0]
         CVa = PowerExpand[√avar / abar]
```

Out[7]= $\mathtt{gbar}^2$

Out[8]= $3 \, \mathtt{gbar}^4$

Out[9]= $\sqrt{3}$

Given that we are modelling pores as rectangles, the expression obtained for the mean seems intuitively reasonable. We observe that the coefficient of variation is independent of the mean. Now, if a random variable x is non-negative, *i.e.* $x \geq 0$, and has coefficient of variation independent of the mean, then this implies that x is distributed according to a gamma distribution. Comparison of the probability density we have derived (solid line) with that of a gamma distribution with the same mean and variance (dashed line) suggests that a gamma distribution would provide a reasonable approximation for the probability density of polygon areas:

In[10]:= **Plot[{pdfa /. gbar → 1, PDF[GammaDistribution[1 / 3, 3], a]},**
{a, 0, 5}, PlotRange → {0, 2},
PlotStyle → {{}, Dashed}, AxesLabel → {"a", "g(a)"}]

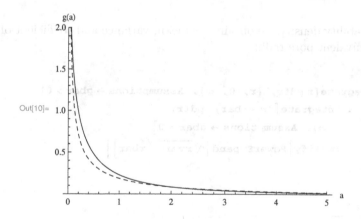

Out[10]=

Note also that the probability density we have obtained resembles the Monte Carlo data of Crain and Miles [22] for random polygons.

When describing fibrous filters and general classes of stochastic porous materials, it is often convenient to characterise voids by some equivalent pore diameter. A widely used descriptor is an equivalent diameter based on the hydraulic radius, r_h of a pore as given by the ratio of the pore area to its perimeter:

$$r_h = \frac{a}{p} \, . \tag{3.23}$$

A circle of unit radius has hydraulic radius, $r_h = 1/2$, so the equivalent diameter of voids with arbitrary shape is given by

$$d_{eq} = 4\,r_h \ . \tag{3.24}$$

Although a probability density of hydraulic radii or equivalent diameters would be an attractive thing to have, any derivation would require knowledge of the joint probability of pore areas and perimeters. In the absence of the required joint probability density, Corte and Lloyd [20] derived the probability density of the radii of circles with the same areas as their rectangular voids, i.e.

$$r = \sqrt{\frac{a}{\pi}} \ , \tag{3.25}$$

such that the void perimeter distribution is not required and the probability density of r is obtained by a variable transform:

In[11]:= **a = $\pi\,$r^2;**
    pdfr = PowerExpand[D[a, r] pdfa]

Out[12]= $\dfrac{4\,\pi\,r\,\mathrm{BesselK}\!\left[0,\ \frac{2\sqrt{\pi}\,r}{gbar}\right]}{gbar^2}$

From this probability density, we obtain the mean, variance and coefficient of variation of equivalent pore radii:

In[13]:= **rbar =**
    Integrate[r pdfr, {r, 0, ∞}, Assumptions $\to$ gbar > 0]
    rvar = Integrate$\left[$(r - rbar)^2 pdfr,
    {r, 0, ∞}, Assumptions $\to$ gbar > 0$\right]$
    cvr = Simplify$\left[$PowerExpand$\left[\sqrt{\mathrm{rvar}}\ /\ \mathrm{rbar}\right]\right]$
    N[cvr]

Out[13]= $\dfrac{gbar\,\sqrt{\pi}}{4}$

Out[14]= $gbar^2\left(\dfrac{1}{\pi} - \dfrac{\pi}{16}\right)$

Out[15]= $\dfrac{\sqrt{16 - \pi^2}}{\pi}$

Out[16]= 0.788124

Again, we observe that the coefficient of variation is independent of the mean and accordingly we find that the probability density of equivalent pore radii (solid line) is well approximated by a gamma distribution with the same mean and coefficient of variation (dashed line):

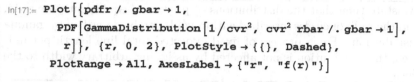

```
In[17]:= Plot[{pdfr /. gbar → 1,
          PDF[GammaDistribution[1/cvr², cvr² rbar /. gbar → 1],
          r]}, {r, 0, 2}, PlotStyle → {{}, Dashed},
        PlotRange → All, AxesLabel → {"r", "f(r)"}]
```

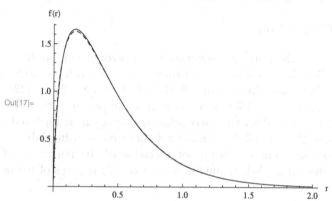

From the treatment of Corte and Lloyd [20] we expect therefore that experimental measurements of the pore radius distribution should be quite well described by the gamma distribution and a plot of the standard deviation against the mean should be linear and pass through the origin such that its gradient is the coefficient of variation of the distribution. There is strong experimental evidence spanning several decades indicating that this is the case for near-planar materials such as paper [9, 20, 35, 135, 141]. Such measurements do include a contribution from out-of-plane pores [146], though the same behaviour is reported for thin fibre networks where such pores are not an issue [47, 106]. We observe also that from the approximation, the mean equivalent pore radius is proportional to the mean inter-crossing distance which in turn depends on intensity of the lines process; this seems intuitively reasonable and it is consistent with the observations of Castro and Ostoja-Starzewski [13] in their study of stochastic sieving by fibrous filters. Finally, we mention a special case considered by Miles [102]. He showed that the probability density of the radii of circles inscribed in random polygons, *i.e.* the largest circle contained in a convex polygon that touches at least three of its sides, is exponential with mean equivalent to the mean inter-crossing distance; so again we find the characteristic dimension of a polygonal void is distributed according to a gamma distribution. Importantly, Miles showed also that the distribution of polygon areas and inscribed circle radii is unaffected by adding width to the lines; the smallest polygons being closed by adding width and larger polygons becoming smaller, this holding for arbitrary distributions of line width [102].

> We state then that the distributions of pore area, equivalent pore radii, and the radii of circles inscribed within voids can be described by gamma distributions. For a process of fixed expected total line length per unit area, the distribution is not affected by assigning arbitrary width to the lines.

3.3.4 Intrinsic Correlation

We have observed that the polygons generated by isotropic random line processes are convex. Graphical realisations of such processes, such as the Monte Carlo studies of Piekaar and Clarenburg [16, 17, 124], Crain and Miles [22] and Castro and Ostoja-Starzewski [13] tend to yield random polygons that appear 'roundish', nearly regular, rather than irregular in shape; this is reported also by Corte and Lloyd [20] for cellulosic fibre networks made using laboratory and industrial processes. The coefficient of variation of the perimeter of n-sided polygons is that of a χ^2-distribution with $2(n-2)$ degrees of freedom, *i.e.*

$$cv(p, n) = \frac{1}{\sqrt{n-2}}$$

and therefore decreases with increasing n, which is consistent with polygons becoming more regular as the number of sides, perimeter and area increase. Robust proof that regularity is a limiting property for random polygons as their area, perimeter or number of sides become large is provided by Miles [103] and Kovalenko [89].

An important implication of the 'roundness' or regularity of polygons is that the lengths of their sides are not independent, as assumed by Corte and Lloyd [20], but are to some extent correlated [41]. Qualitative evidence for such correlation can be obtained by inspection of a graphical representation of a random line process such as that generated on page 72. This reveals that in regions of high density there are many short inter-crossing distances, and that in regions of low density there are fewer but longer inter-crossing distances. So we have positive correlation between nearby polygonal side lengths and this tends to yield more regular polygons, simply from the random variations in the local density that arise from the underlying Poisson line process: evidently random isotropy has an inherent correlation of adjacent inter-crossing distances, and this gives rise to polygonal voids that seem mainly 'roundish' in real materials.

The degree of correlation between adjacent polygon sides is not known analytically, though it has recently been determined numerically [41]. Here we develop code for a Monte-Carlo approach to the problem. When developing such code, it is good practice to begin with a small sample so that the size

of the output is limited, aiding its inspection. We begin then by generating
the equations of random lines passing through a unit square and consider just
five lines in the first instance.

```
In[1]:= nlines = 5; nseed = 1;
        SeedRandom[nseed]
        xi = RandomReal[{0, 1}, {nlines}];
        SeedRandom[nseed + 1]
        yi = RandomReal[{0, 1}, {nlines}];
        SeedRandom[nseed + 2]
        θi = RandomReal[{-π / 2, π / 2}, {nlines}];
        lines = Tan[θi] (x - xi) + yi;
```

The repeated use of **SeedRandom** is included to allow the evolution of correlation to be examined. So if we change **nlines** to 6, then we obtain the
original five lines and one other.

The polygon sides of interest are generated at the points of intersection of
the random lines and each line intersects all lines except itself. Some of these
intersections will occur outside the unit square and we will remove these from
our analysis at a subsequent stage of our implementation. We compute the
x-coordinates of the points of intersection for each line such that each sublist
of the output is associated with the equation of a line in the same location of
the list **lines**:

```
In[9]:= X = Table[x /. Solve[lines[[i]] == lines[[j]], x][[1]],
        {i, 1, nlines - 1}, {j, i + 1, nlines}]
```

```
Out[9]= {{0.0933374, 0.231166, 0.0771163, 0.122874},
         {0.0916032, 0.0943121, 0.0920709},
         {0.033859, 0.0759685}, {-0.0437588}}
```

Paired (x, y) coordinates of the crossing points are obtained by substituting
these x-coordinates into the corresponding equation of a line:

```
In[10]:= XY =
         Table[Transpose[{X[[i]], lines[[i]] /. x → X[[i]]}],
         {i, 1, nlines - 1}]
```

```
Out[10]= {{{0.0933374, 0.771097}, {0.231166, 0.761797},
          {0.0771163, 0.772191}, {0.122874, 0.769104}},
         {{0.0916032, 0.834556}, {0.0943121, 0.735433},
          {0.0920709, 0.81744}},
         {{0.033859, 0.86466}, {0.0759685, 0.842707}},
         {{-0.0437588, 1.03058}}}
```

We seek to associate each of these coordinates with those of the adjacent crossings on the pair of lines generating it and accordingly, we associate each coordinate with an identifier. Our subsequent code takes advantage of *Mathematica*'s advanced list handling capabilities, but it is simplified if each crossing is included twice, such that a crossing generated by lines i and j is associated with the identifiers {**i, j**} and {**j, i**}.

```
In[11]:= sourcelines = Drop[ Table[{i, j},
            {i, 1, nlines}, {j, i + 1, nlines}], -1];
        XYid1 = Table[{sourcelines[[i, j]], XY[[i, j]]},
            {i, 1, nlines - 1}, {j, 1, Length[XY[[i]]]}];
        XYid2 = Table[{Reverse[sourcelines[[i, j]]], XY[[i, j]]},
            {i, 1, nlines - 1}, {j, 1, Length[XY[[i]]]}];
        XYid = Partition[Sort[Partition[Partition[
            Flatten[Join[XYid1, XYid2]], 2], 2]], nlines - 1]
```

```
Out[14]= {{{{1, 2}, {0.0933374, 0.771097}}, {{1, 3}, {0.231166, 0.761797}},
          {{1, 4}, {0.0771163, 0.772191}}, {{1, 5}, {0.122874, 0.769104}}},
         {{{2, 1}, {0.0933374, 0.771097}}, {{2, 3}, {0.0916032, 0.834556}},
          {{2, 4}, {0.0943121, 0.735433}}, {{2, 5}, {0.0920709, 0.81744}}},
         {{{3, 1}, {0.231166, 0.761797}}, {{3, 2}, {0.0916032, 0.834556}},
          {{3, 4}, {0.033859, 0.86466}}, {{3, 5}, {0.0759685, 0.842707}}},
         {{{4, 1}, {0.0771163, 0.772191}}, {{4, 2}, {0.0943121, 0.735433}},
          {{4, 3}, {0.033859, 0.86466}}, {{4, 5}, {-0.0437588, 1.03058}}},
         {{{5, 1}, {0.122874, 0.769104}}, {{5, 2}, {0.0920709, 0.81744}},
          {{5, 3}, {0.0759685, 0.842707}}, {{5, 4}, {-0.0437588, 1.03058}}}}
```

We replace any crossings occurring outside the unit square with an empty sublist. These are removed using **Flatten** and the original list structure is recovered using **Partition**.

```
In[15]:= XYid = Table[Table[If[(0 ≤ XYid[[i, j, 2, 1]]) ≤ 1. &&
            0 ≤ XYid[[i, j, 2, 2]]) ≤ 1.), XYid[[i, j]], {}],
            {j, 1, Length[XYid[[i]]]}], {i, 1, nlines}];
        XYid = Table[Partition[Partition[Flatten[XYid[[i]]],
            2], 2], {i, 1, Length[XYid]}]
```

```
Out[16]= {{{{1, 2}, {0.0933374, 0.771097}}, {{1, 3}, {0.231166, 0.761797}},
          {{1, 4}, {0.0771163, 0.772191}}, {{1, 5}, {0.122874, 0.769104}}},
         {{{2, 1}, {0.0933374, 0.771097}}, {{2, 3}, {0.0916032, 0.834556}},
          {{2, 4}, {0.0943121, 0.735433}}, {{2, 5}, {0.0920709, 0.81744}}},
         {{{3, 1}, {0.231166, 0.761797}}, {{3, 2}, {0.0916032, 0.834556}},
          {{3, 4}, {0.033859, 0.86466}}, {{3, 5}, {0.0759685, 0.842707}}},
         {{{4, 1}, {0.0771163, 0.772191}}, {{4, 2}, {0.0943121, 0.735433}},
          {{4, 3}, {0.033859, 0.86466}}}, {{{5, 1}, {0.122874, 0.769104}},
          {{5, 2}, {0.0920709, 0.81744}}, {{5, 3}, {0.0759685, 0.842707}}}}
```

To identify adjacent crossings on a given line, we sort each sublist by the
x-coordinate of the crossings and to identify polygon sides that extend
beyond the unit square we append the dummy coordinate and identifier
{{0, 0},{0, 0}} to each sublist:

```
In[17]:= XYid = Table[Sort[Table[Reverse[XYid[[j, i]]], {i, 1,
            Length[XYid[[j]]]}]], {j, 1, Length[XYid]}];
         XYid = Table[Append[Prepend[XYid[[i]],
            {{0, 0}, {0, 0}}], {{0, 0}, {0, 0}}], {i, 1, nlines}]
```

```
Out[18]= {{{{0, 0}, {0, 0}}, {{0.0771163, 0.772191}, {1, 4}},
         {{0.0933374, 0.771097}, {1, 2}}, {{0.122874, 0.769104}, {1, 5}},
         {{0.231166, 0.761797}, {1, 3}}, {{0, 0}, {0, 0}}},
         {{{0, 0}, {0, 0}}, {{0.0916032, 0.834556}, {2, 3}},
         {{0.0920709, 0.81744}, {2, 5}}, {{0.0933374, 0.771097}, {2, 1}},
         {{0.0943121, 0.735433}, {2, 4}}, {{0, 0}, {0, 0}}},
         {{{0, 0}, {0, 0}}, {{0.033859, 0.86466}, {3, 4}},
         {{0.0759685, 0.842707}, {3, 5}}, {{0.0916032, 0.834556}, {3, 2}},
         {{0.231166, 0.761797}, {3, 1}}, {{0, 0}, {0, 0}}},
         {{{0, 0}, {0, 0}}, {{0.033859, 0.86466}, {4, 3}},
         {{0.0771163, 0.772191}, {4, 1}}, {{0.0943121, 0.735433}, {4, 2}},
         {{0, 0}, {0, 0}}}, {{{0, 0}, {0, 0}},
         {{0.0759685, 0.842707}, {5, 3}}, {{0.0920709, 0.81744}, {5, 2}},
         {{0.122874, 0.769104}, {5, 1}}, {{0, 0}, {0, 0}}}}
```

Since each intersection generates four polygon sides, we need to extract
groups of five coordinates where the first coordinate represents an intersection and the other four coordinates represent the end points of the polygon
sides associated with it. The following line of code achieves this by selecting
each coordinate along a line and those either side of it; the coordinates of the
adjacent crossings on the line intersecting the given line are obtained using
Position to identify the location of these crossing points in the sublist associated with the intersecting line.

```
In[19]:= G5 = Table[{XYid[[i, j, 1]],
            XYid[[i, j - 1, 1]], XYid[[i, j + 1, 1]],
            XYid[[XYid[[i, j, 2, 2]]]][[Position[
                XYid[[XYid[[i, j, 2, 2]]]],
                XYid[[i, j, 2, 1]]][[1, 1]] - 1]][[1]],
            XYid[[XYid[[i, j, 2, 2]]]][[Position[
                XYid[[XYid[[i, j, 2, 2]]]],
                XYid[[i, j, 2, 1]]][[1, 1]] + 1]][[1]]},
         {i, 1, nlines}, {j, 2, Length[XYid[[i]]] - 1}];
```

The output of **G5** is rather large, even for a system of just five lines. Inspection
of the first sublist reveals the groups of five coordinates defining the polygon

sides originating from each point of intersection along line 1. We bear in mind that the coordinate {0, 0} indicates that a polygon side extends beyond the unit square and must be excluded from the analysis.

In[20]:= **G5[[1]]**

Out[20]= {{{0.0771163, 0.772191}, {0, 0}, {0.0933374, 0.771097},
        {0.033859, 0.86466}, {0.0943121, 0.735433}},
      {{0.0933374, 0.771097}, {0.0771163, 0.772191}, {0.122874, 0.769104},
        {0.0920709, 0.81744}, {0.0943121, 0.735433}},
      {{0.122874, 0.769104}, {0.0933374, 0.771097}, {0.231166, 0.761797},
        {0.0920709, 0.81744}, {0, 0}}, {{0.231166, 0.761797},
        {0.122874, 0.769104}, {0, 0}, {0.0916032, 0.834556}, {0, 0}}}

To compute the lengths of the polygon sides, we apply Pythagoras' theorem and assign a side length of zero to any sides extending beyond the unit square. The output is a list of polygon sides where the first two elements are associated with a given line and the second two elements are associated with the line that intersects it.

In[21]:= **G5 = Flatten[G5, 1];**
       sides = Table[Table[If[G5[[i, j]] ≠ {0, 0},
          Sqrt[(G5[[i, 1, 1]] - G5[[i, j, 1]])^2 +
            (G5[[i, 1, 2]] - G5[[i, j, 2]])^2], 0],
          {j, 2, 5}], {i, 1, Length[G5]}];
       sides[[1]]

Out[23]= {0, 0.0162581, 0.102087, 0.040582}

We proceed to pair each polygon side with those adjacent to it and remove all instances of polygon sides extending beyond the unit square:

In[24]:= **pairs = Flatten[Table[{{sides[[i, 1]], sides[[i, 3]]},**
          {sides[[i, 3]], sides[[i, 2]]}, {sides[[i, 2]],
          sides[[i, 4]]}, {sides[[i, 4]], sides[[i, 1]]}},
          {i, 1, Length[sides]}], 1];
       pairs = Cases[pairs, {x_, y_} /; y > 0 /; x > 0];

The list **pairs** includes each pair of polygon sides twice, since we have generated our pairs by inspecting the intersections for each line. To remove duplicate entries, we sort each element of our list and then sort the resulting list by the first element of each sublist; finally we remove alternate entries. The resultant output is a list of pairs of polygon sides, sorted such that the shorter side is the first of the pair.

```
In[26]:= pairs = Sort[
             Table[Sort[pairs[[i]]], {i, 1, Length[pairs]}]];
         pairs = Take[pairs, {1, Length[pairs], 2}];
```

This sorting is, in fact, a very important stage of our analysis. Without it, we would necessarily have no correlation. To compute the correlation, as defined on page 50, we separate our sorted pairs into two lists and use the command **Correlation**.

```
In[28]:= pairsXY = Transpose[pairs];
         X = pairsXY[[1]];
         Y = pairsXY[[2]];
         Correlation[X, Y]
```

```
Out[31]= 0.584382
```

By enclosing all of our code in brackets and using a **SetDelayed** operator we can assign the full code we have written to a single command that sets the number of lines and the random seed; the collected code is shown on page 100. We may now readily compute, for example, the influence of different random seeds on the correlation between adjacent polygon sides generated by 50 random lines in the unit square:

```
In[33]:= Table[ComputeCorr[50, i], {i, 1, 10}]
```

```
Out[33]= {0.598026, 0.607214, 0.616877, 0.640122, 0.60286,
          0.625369, 0.62703, 0.589316, 0.593109, 0.601946}
```

This is good practice, since for our Monte Carlo method to produce robust data, we require that the outputs are not overly sensitive to the choice of random seed.

Before considering the correlation further, it is worthwhile to check that our code generates data which can be considered to faithfully represent the system of interest. A good test for our example is whether the distribution of polygon sides is exponentially distributed. We check this for a network of 200 lines by computing the coefficient of variation of side lengths, which should be close to 1:

```
In[34]:= ComputeCorr[200, 1];
         sides = DeleteCases[Flatten[sides], 0];
         m = Mean[sides];
         StandardDeviation[sides] / m
```

```
In[32]:= ComputeCorr[nl_, ns_] :=
         (nlines = nl; nseed = ns; SeedRandom[nseed];
          xi = RandomReal[{0, 1}, {nlines}]; SeedRandom[nseed + 1];
          yi = RandomReal[{0, 1}, {nlines}]; SeedRandom[nseed + 2];
          θi = RandomReal[{-π / 2, π / 2}, {nlines}];
          lines = Tan[θi] (x - xi) + yi;
          X = Table[x /. Solve[lines[[i]] == lines[[j]], x][[1]],
            {i, 1, nlines - 1}, {j, i + 1, nlines}];
          XY = Table[Transpose[{X[[i]], lines[[i]] /. x → X[[i]]}],
            {i, 1, nlines - 1}]; sourcelines = Drop[
           Table[{i, j}, {i, 1, nlines}, {j, i + 1, nlines}], -1];
          XYid1 = Table[{sourcelines[[i, j]], XY[[i, j]]},
            {i, 1, nlines - 1}, {j, 1, Length[XY[[i]]]}];
          XYid2 = Table[{Reverse[sourcelines[[i, j]]], XY[[i, j]]},
            {i, 1, nlines - 1}, {j, 1, Length[XY[[i]]]}];
          XYid = Partition[Sort[Partition[Partition[
              Flatten[Join[XYid1, XYid2]], 2], 2]], nlines - 1];
          XYid = Table[Table[If[(0 ≤ XYid[[i, j, 2, 1]] ≤ 1. &&
               0 ≤ XYid[[i, j, 2, 2]] ≤ 1.), XYid[[i, j]], {}],
            {j, 1, Length[XYid[[i]]]}], {i, 1, nlines}];
          XYid = Table[Partition[Partition[Flatten[XYid[[i]]], 2],
            2], {i, 1, Length[XYid]}];
          XYid = Table[Sort[Table[Reverse[XYid[[j, i]]],
              {i, 1, Length[XYid[[j]]]}]], {j, 1, Length[XYid]}];
          XYid = Table[Append[Prepend[XYid[[i]], {{0, 0}, {0, 0}}],
            {{0, 0}, {0, 0}}], {i, 1, nlines}];
          G5 = Table[{XYid[[i, j, 1]], XYid[[i, j - 1, 1]],
             XYid[[i, j + 1, 1]], XYid[[XYid[[i, j, 2, 2]]]][[
               Position[XYid[[XYid[[i, j, 2, 2]]]],
                 XYid[[i, j, 2, 1]]][[1, 1]] - 1]][[1]],
             XYid[[XYid[[i, j, 2, 2]]]][[Position[XYid[[XYid[[i, j, 2,
                  2]]]], XYid[[i, j, 2, 1]]][[1, 1]] + 1]][[1]]},
            {i, 1, nlines}, {j, 2, Length[XYid[[i]]] - 1}];
          G5 = Flatten[G5, 1]; sides =
           Table[Table[If[G5[[i, j]] ≠ {0, 0},
             Sqrt[(G5[[i, 1, 1]] - G5[[i, j, 1]])^2 +
               (G5[[i, 1, 2]] - G5[[i, j, 2]])^2], 0],
             {j, 2, 5}], {i, 1, Length[G5]}]; pairs =
           Flatten[Table[{{sides[[i, 1]], sides[[i, 3]]},
              {sides[[i, 3]], sides[[i, 2]]},
              {sides[[i, 2]], sides[[i, 4]]}, {sides[[i, 4]],
               sides[[i, 1]]}}, {i, 1, Length[sides]}], 1];
          pairs = Cases[pairs, {x_, y_} /; y > 0 /; x > 0];
          pairs =
           Sort[Table[Sort[pairs[[i]]], {i, 1, Length[pairs]}]];
          pairs = Take[pairs, {1, Length[pairs], 2}];
          pairsXY = Transpose[pairs];
          X = pairsXY[[1]];
          Y = pairsXY[[2]];
          Correlation[X, Y])
```

Out[37]= 1.0132

Secondly, a histogram of the side length distribution agrees well with the probability density of an exponential distribution with the same mean:

```
In[38]:= Needs["Histograms`"]
        Show[Histogram[sides, HistogramScale → 1,
          HistogramCategories → 50], Plot[
          PDF[ExponentialDistribution[1/m], x], {x, 0, .1},
          PlotStyle → Thickness[0.005], PlotRange → All]]
```

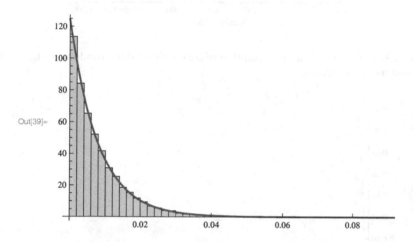

Importantly, discarding the polygon sides that extend beyond the unit square from our analysis has had no significant influence on the distribution of polygon sides, since these are exponential, as expected. Given this, we can be confident that the polygon side distribution used to calculate correlation can be considered representative of the population of polygon sides.

We track the evolution of a stable correlation by evaluating **ComputeCorr** for increasing numbers of lines:

```
In[40]:= nl = {5, 6, 7, 8, 9, 10, 12, 14, 16, 18,
          20, 30, 40, 50, 75, 100, 125, 150, 200, 500};
        evolve = Transpose[{nl, Table[ComputeCorr[nl[[i]], 1],
            {i, 1, Length[nl]}]}];
        ListLinePlot[evolve]
```

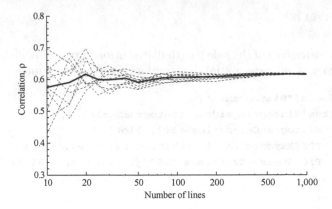

Figure 3.8. Collected results for evolution of correlation for ten random seeds. Solid line shows mean correlation

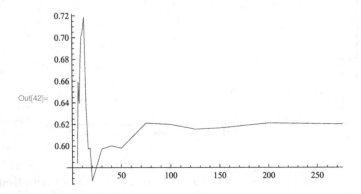

Collected results for the correlation between adjacent polygon sides as calculated by evaluating **ComputeCorr** for increasing numbers of lines generated with ten different random seeds are shown in Figure 3.8. We observe that these converge very rapidly on a stable value of the correlation which we compute to be $\rho = 0.616 \pm 0.001$ [41].

Now, we have already noted that to some extent, correlation arises directly from the process of sorting our random variables. We therefore require a value for the correlation between sorted but *independent* pairs of exponentially distributed random variables against which we may compare the correlation which we have computed for adjacent polygon sides.

Sorted Independent Polygon Sides

To determine the correlation between sorted independent pairs of exponentially distributed random variables (x, y) with $x \le y$, we require first distributions of x and y. The joint distribution of $X = x \cup y$ is exponential, and we consider the case with unit mean such that the probability density of X is

$$f(X) = e^{-X} . \tag{3.26}$$

To guide our derivation of the probability densities of x and y, it is helpful to consider the process of selecting pairs (x, y) from the source distribution. First, we select a random number from the source exponential distribution; at this stage it is unknown whether this random number, X, will form part of the population of x or of y. Secondly, we select a second random number from the source distribution. If this number is less than the first, then it forms part of the population of x; otherwise it forms part of the population of y. This step represents the sorting process. The probability that the second number is less than X is

$$F(X) = \int_0^X f(X) \, \mathrm{d}X . \tag{3.27}$$

Inevitably, $0 \le F(X) \le 1$ and exists for all probability densities, $f(X)$ with limits of integration appropriate to the domain of the random variable. Such functions are called cumulative distribution functions.

Returning to our sorted exponentially distributed variables, the probability that the second number is less than the first is $F(X)$ and the probability that it is greater than the first is $(1 - F(X))$. The marginal probabilities of x and y are therefore

$$p(x) = (1 - F(x) + (1 - F(x))) \, f(x) = 2 \, (1 - F(x)) \, f(x) \tag{3.28}$$
$$q(y) = (1 + F(y) - (1 - F(y))) \, f(y) = 2 \, F(y) \, f(y) , \tag{3.29}$$

such that $p(X) + q(X) = f(X)$.

Recalling Equations 2.32 and 2.35, and bearing in mind that the expected value of the product, $\overline{xy} = 1$ is unaltered by sorting, the correlation is given by

$$\rho = \frac{\overline{xy} - \bar{x}\,\bar{y}}{\sigma(x)\,\sigma(y)} \tag{3.30}$$

$$= \frac{1 - \bar{x}\,\bar{y}}{\sigma(x)\,\sigma(y)} . \tag{3.31}$$

so we must carry out the appropriate integrations to determine the mean and variance of x and y in order that we may determine ρ.

```
In[43]:=  pdfx = Refine[2 (1 - CDF[ExponentialDistribution[1], x])
              PDF[ExponentialDistribution[1], x], x > 0]
          pdfy = Refine[2 (CDF[ExponentialDistribution[1], y])
              PDF[ExponentialDistribution[1], y], y > 0]
```

Out[43]= $2 \, e^{-2 \, x}$

Out[44]= $2 \, e^{-y} \, (1 - e^{-y})$

In[45]:= $\{$ **xbar = Integrate[x pdfx, {x, 0, ∞}],**
    xvar = Integrate$\left[\text{(x - xbar)}^2 \text{pdfx, {x, 0, ∞}} \right]$,
    ybar = Integrate[y pdfy, {y, 0, ∞}],
    yvar = Integrate$\left[\text{(y - ybar)}^2 \text{pdfy, {y, 0, ∞}} \right] \}$

Out[45]= $\left\{ \dfrac{1}{2}, \dfrac{1}{4}, \dfrac{3}{2}, \dfrac{5}{4} \right\}$

In[46]:= **covariance = 1 - xbar ybar**

Out[46]= $\dfrac{1}{4}$

In[47]:= **ρ = covariance $\Big/ \sqrt{\text{xvar yvar}}$**
    N[ρ]

Out[47]= $\dfrac{1}{\sqrt{5}}$

Out[48]= 0.447214

From this treatment, we are able to state that the correlation determined for sorted adjacent polygon sides is greater than that arising from the sorting of independent polygon sides. Referring to Figure 3.8, we observe also that for processes of 20 or more lines per unit area, the correlation is always greater than that calculated for independent and ordered pairs.

We state then that the lengths of the adjacent sides of polygons generated by a random line process in the plane are correlated. This correlation is consistent with the observed 'roundness' of polygonal voids.

4

Poisson Fibre Processes I: Fibre Phase

4.1 Introduction

So far, we have considered Poisson processes of points and of lines with no width and infinite length. Of course, real fibres have finite length and width, so they occupy space in the plane of the network and generate contacts only if their proximity and orientation relative to each other fulfil certain criteria. Over the next two chapters we will consider some statistics of random networks of fibres modelled as rectangles with finite length and width. We begin by considering thin networks that may be considered to exist in two dimensions only and proceed to develop models for materials with finite thickness that can be represented as multi-planar structures consisting of several layers of thin structures. In this chapter we concentrate on interactions between fibres and the distribution of mass in the plane of the network. We extend our analysis to consider the porous structure of random near-planar fibrous materials in Chapter 5.

4.2 Planar Fibre Networks

A classical problem of relevance is known as Buffon's Needle problem, first posed in 1773 by Georges-Louis Leclerc, Comte de Buffon [5]. The problem seeks the probability that a needle of length λ intersects a line if dropped at random on a plane marked with equally spaced parallel lines separated by a distance, d where $d > \lambda$. It is illustrated graphically in Figure 4.1.

Our random variables are the distance of the centre of the needle from the nearest line, and the orientation of the needle to, say, the direction of the lines; we denote these uniformly distributed random variables x and θ, respectively. Taking advantage of symmetry, we define $0 \leq x \leq d/2$ and $0 \leq \theta \leq \pi/2$; we input these probability densities to *Mathematica* as

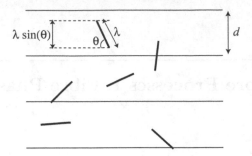

Figure 4.1. Buffon's Needle problem

In[1]:= **pdfx = 2 / d;**
pdfθ = 2 / π;

To obtain the probability of intersecting a line, we must integrate the product of these probability densities, first with respect to x and then with respect to θ. Note that when evaluating multiple integrals in *Mathematica*, the first variable given corresponds to the outermost integral, and is done last. The probability that a needle intersects a line is therefore

In[3]:= **Integrate[pdfx pdfθ, {θ, 0, π / 2}, {x, 0, λ Sin[θ] / 2}]**

Out[3]= $\dfrac{2\,\lambda}{d\,\pi}$

So far, we have considered networks of lines assuming that they have no width. A convenient property of lines is that their intersections occur at points and they stochastically subdivide the plane without occupying any space. Accordingly, we could state with confidence that no set of three lines intersects at a point and that the sum of the areas of all polygons within a given region is the same as the area of that region. These statements no longer hold once we include width and transform our lines into rectangles so a given point in space may be covered by more than two rectangles and in a random fibre process with fixed expected fibre length per unit area, the fraction of the area covered by fibres will increase with increasing width.

A convenient random variable to characterise our fibre process in the plane is the *coverage*, which we define as the number of fibres covering points in the plane of support of the network. We have seen that the number of fibre centres and the number of fibre crossings are distributed according to a point Poisson process in the plane and this is the same for the discrete random variable coverage also. Recall that the Poisson distribution is fully characterised by its mean, so the probability that a point has coverage c in a network of mean coverage $\bar{c}$ is

$$P(c) = \frac{\bar{c}^c\, e^{-\bar{c}}}{c!} \qquad \text{for } c = 0, 1, 2, 3\dots \tag{4.1}$$

The mean coverage $\bar{c}$ is related to the expected number of fibres per unit area μ by

$$\bar{c} = \mu\,\lambda\,\omega \tag{4.2}$$

Of course, the expected number of fibres per unit area is not typically known. Accordingly, the mean coverage is typically calculated from the mean mass per unit area, or areal density, of the network, $\bar{\beta}$, the fibre width ω and the mass per unit length of fibres, or linear density δ. The mass per unit projected area of the fibre is

$$\beta_f = \frac{\delta}{\omega}\,, \tag{4.3}$$

and it follows that the expected coverage is given by

$$\bar{c} = \frac{\bar{\beta}}{\beta_f} = \frac{\omega\,\bar{\beta}}{\delta}\,. \tag{4.4}$$

Many synthetic fibres have circular cross section and are formed from materials with known density ρ_f and the mean coverage of networks of these fibres is

$$\bar{c} = \frac{4\,\bar{\beta}}{\pi\,\omega\,\rho_f}\,. \tag{4.5}$$

We are immediately able to make some statements about coverage. As a Poisson random variable coverage may take only integer values, the probability of integer coverage c is influenced by the mean coverage only and it is therefore proportional to the areal density of the network and influenced by fibre width and coarseness. If fibre width is independent of coarseness, as might be the case for hollow fibres, including many natural fibres, then the mean coverage is directly proportional to fibre width and it is inversely proportional to coarseness. For networks of solid fibres with circular cross section, the mean coverage is proportional to the mean areal density of the network, but now is inversely proportional to fibre width. This apparent contradiction arises because, in the solid-fibre case, if we increase fibre width, we increase also the mass of the fibre, $\lambda\,\delta$ and so require fewer fibres per unit area to provide a given areal density. Note that the intensity of the Poisson fibre process, $i.e.$ the total fibre length per unit area is

$$\tau_f = \frac{\bar{c}}{\omega} \tag{4.6}$$

$$= \frac{\bar{\beta}}{\delta}\,, \tag{4.7}$$

so we observe that the mean coverage and the total fibre length per unit area are independent of fibre length.

A random network of rectangles can be considered as a projection of a fibre network onto a plane. This can be useful when considering for example,

the distribution of fibre mass in the plane or the likelihood of a pinhole, *i.e.*
a region covered by no fibres. The probability of pinholes in a network with
coverage $\bar{c}$ is, of course, the fraction of the network with coverage zero, *i.e.*

$$\epsilon = P(0) = e^{-\bar{c}} . \tag{4.8}$$

Now, whereas we may use Poisson statistics of coverage to obtain expressions
for properties that may be represented by their projection onto a plane, models for fibre contact are complicated by the fact that in regions with coverage
$c > 1$, vertically adjacent fibres may or may not make contact, depending on
the influence of nearby fibres. Inevitably, the likelihood of vertically adjacent
fibres being separated by some distance increases with increasing coverage and
network porosity. We will consider models that account for the out-of-plane
structure of the network in Section 4.3 but consider first the thinnest fibre
networks for which truly planar models can be applied. Such networks are
often classified as being 'two-dimensional' and it is assumed that wherever
two fibres cross, they generate a contact. For this criterion to be met, Kallmes
and Corte defined two-dimensional networks as those where, "...the number
of fibres in the network is so small that the area covered by more than two
fibres is negligible, *i.e.* less than 1 %." [74]. The probability that a network
has coverage greater than 2 is

In[4]:= **P[c_] := PDF[PoissonDistribution[cbar], c]**
       Pgt2 = FullSimplify[1 - (P[0] + P[1] + P[2])]

Out[5]= $1 - \dfrac{1}{2} \, (2 + \text{cbar} \, (2 + \text{cbar})) \, e^{-\text{cbar}}$

We use **FindRoot** to identify the maximum coverage for which this fraction
is less than 1 %:

In[6]:= **FindRoot[Pgt2 == (1 / 100), {cbar, 1}]**

Out[6]= {cbar → 0.436045}

So using Kallmes and Corte's definition of a two-dimensional random fibre
network, we have an upper bound on mean coverage of $\bar{c} \approx 0.44$. Quite often,
theory for two-dimensional networks is considered to apply if $\bar{c} \leq 1$ and for
such networks, the probability of coverage *three* is less than 2 %:

In[7]:= **1 - (P[0] + P[1] + P[2] + P[3]) /. cbar → 1.**

Out[7]= 0.0189882

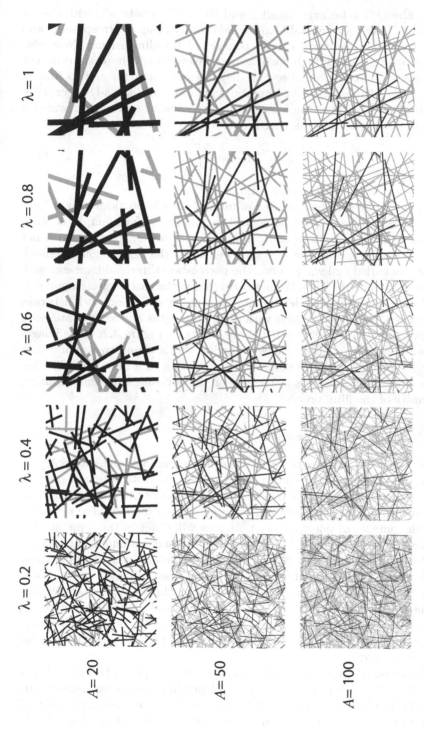

Figure 4.2. Networks of fibres with length λ and aspect ratio, A at the limiting mean coverage to be considered two-dimensional, *i.e.* with $\bar{c} = 0.436$; fibres shown in black represent networks at the percolation threshold, as given by Equation 4.9

For most fibres, β_f is between about 5 and 10 g m^{-2} so networks with similar areal density, which are realisable in the laboratory using electrospinning and wet- or dry-laying processes, can be considered two-dimensional. Note also that there exists a lower limit for coverage below which the network is not fully connected, but consists of several disjointed sub-structures. This limit is termed the *percolation threshold*; from the simulations of Pike and Seager [125], Niskanen *et al.* [108] give the lower limit of the mean coverage at which a network of fibres with uniform length and width percolates as

$$\bar{c}_{\text{perc}} = 5.7 \frac{\omega}{\lambda} = \frac{5.7}{A} , \tag{4.9}$$

where $A = \lambda/\omega$ is the fibre aspect ratio.

The percolation threshold is often of interest since it provides a lower limit on interconnectivity among constituent fibres and hence, for example, network conductivity, see *e.g.* [6, 7]. In many applications however, percolation is not an issue because networks have coverage well above the percolation threshold. Note that for curled and wavy fibres, the percolation threshold increases with increasing non-linearity [168]. We note also that in the case of electrospun networks the fibres may be considered infinitely long; by definition, such networks are percolated. Networks of fibres with different lengths and aspect ratio with mean coverage at the limit $\bar{c} = 0.436$ are shown in Figure 4.2; these images have been rendered for fibre centres occurring within a unit square and with periodic boundary conditions such that any portion of a fibre extending beyond the unit square re-enters the square on the opposing side, thus retaining the intensity of the fibre process.

4.2.1 Probability of Crossing

We have seen that crossings between pairs of random lines generate polygons that can be considered representative of the in-plane distribution of pores in a random fibrous material. We expect the properties of the polygon distribution from such random line networks to apply also to random fibre networks, though the number of polygons per unit area will differ in the same manner as the intensity of crossings. The intensity of crossings is important also because the strength of bonded fibrous networks is determined by the intrinsic strength of the constituent fibres and the number and strength of the bonds between them; bonds occurring in regions of inter-fibre contact. To determine the number of crossings per fibre in a two-dimensional random fibre network, we consider two fibres within an area x^2 of the network; two such fibres are represented by the wide lines in Figure 4.3. We begin by assuming that the fibres have no width.

We compute the probability of intersection between pairs of lines and hence the number of crossings per unit area and per fibre following Kallmes and Corte [74]. The longitudinal axes of the two rectangular fibres in Figure 4.3 cross at an angle θ. The parallelogram surrounding the horizontal line in the

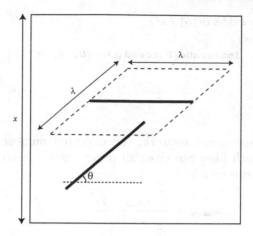

Figure 4.3. Two fibres of length λ oriented at an angle θ to each other within a square area of side x

figure is a rhombus of side λ and for the two lines to cross, the centre of the line with orientation θ must fall within the area of this rhombus. The area of the rhombus is $\lambda^2 \sin(\theta)$ so the probability that these two lines cross is given by the ratio of the area of the rhombus to that of the region of interest, *i.e.*

$$P_{\text{cross},\theta} = \frac{\lambda^2}{x^2} \sin(\theta) \ . \tag{4.10}$$

It follows that the probability density of the angles ϕ between pairs of lines that cross is

$$f(\phi) = \frac{P_{\text{cross},\phi}}{\int_0^{\frac{\pi}{2}} P_{\text{cross},\phi} \, \mathrm{d}\phi}$$
$$= \sin(\phi) \tag{4.11}$$

such that $\bar{\phi} = 1 \approx 57°$.

Now, Equation 4.10 gives the probability that a pair of lines with centres located within an area x^2 and oriented at angle θ to each other will intersect. We seek the probability that *any* pair of lines, with arbitrary orientation, intersect. This is given by

$$\int P_{\text{cross},\theta} \, g(\theta) \, \mathrm{d}\theta \ , \tag{4.12}$$

where $g(\theta)$ is the probability density function for θ. Because of symmetry, we need only consider $0 \le \theta \le \frac{\pi}{2}$. Noting that θ has uniform probability density we have,

In[1]:= **Pcrossθ = λ² Sin[θ] /x²;**
        gθ = 2 / π;
        Pcross = Integrate[Pcrossθ gθ, {θ, 0, π/2}]

Out[3]= $\dfrac{2\,\lambda^2}{\pi\,x^2}$

In our two-dimensional network, the expected number of fibres in an area x^2 is $\bar{n}_{\text{fib}}$. Each fibre can cross all others except itself, so the number of fibre pairs *per unit area* is

$$n_{\text{pairs}} = \frac{\bar{n}_{\text{fib}}(\bar{n}_{\text{fib}} - 1)}{2\,x^2}$$

$$\approx \frac{\bar{n}_{\text{fib}}^2}{2\,x^2} \qquad \text{for } \bar{n}_{\text{fib}} \gg 1 \tag{4.13}$$

where the divisor 2 is applied because each fibre, being one of a pair, is counted twice.

The expected number of fibres per unit area is

$$\bar{n}_{\text{fib}} = \frac{\bar{c}\,x^2}{\lambda\,\omega}, \tag{4.14}$$

so the expected number of crossings *per unit area* is

In[4]:= **npairs = nfib² / (2 x²);**
        nfib = cbar x² / (λ ω);
        ncrosspua = npairs Pcross

Out[6]= $\dfrac{\text{cbar}^2}{\pi\,\omega^2}$

and the number of crossings *per fibre* is

In[7]:= **2 ncrosspua x² / nfib**

Out[7]= $\dfrac{2\,\text{cbar}\,\lambda}{\pi\,\omega}$

where the multiplier 2 is included because each crossing must be counted for each fibre that generates it.

So we have the expected number of crossings per unit area and the expected number of crossings per fibre given by

$$\bar{n}_{\text{cross}} = \frac{\bar{c}^2}{\pi\,\omega^2} \tag{4.15}$$

$$\bar{n}_{\text{cross,fib}} = \frac{2\,\lambda}{\pi\,\omega}\,\bar{c} = \frac{2}{\pi}\,A\,\bar{c} \tag{4.16}$$

respectively, where A is the fibre aspect ratio.

Recalling Equation 4.6, we may express the intensity of crossings in terms of the expected total fibre length per unit area τ_f:

$$\bar{n}_{\text{cross}} = \frac{\tau_f{}^2}{\pi} \tag{4.17}$$

$$\bar{n}_{\text{cross,fib}} = \frac{2\,\tau_f\,\lambda}{\pi} \tag{4.18}$$

> We state then that the number of crossings per unit area depends only on the expected total fibre length per unit area; also, the expected number of crossings per fibre is directly proportional to this parameter and to the length of fibres.

An important implication of our observation that the expected number of crossings per unit area is independent of fibre length is that Equation 4.17 is applicable to networks of fibres with finite or infinite length; the result tells us also that the distributions of inter-crossing distances, and polygon dimensions are independent of fibre length.

An alternative approach to computing the intensity of crossings is provided by Deng and Dodson [27]. Assuming mean coverage $\bar{c} \leq 1$, such that the fraction of the network covered by more than three fibres is negligible, the fraction of the network occupied by fibre crossings is

```
In[8]:= P[c_] := PDF[PoissonDistribution[cbar], c]
        P[2] + 2 P[3]
```

$$\text{Out[9]}= \frac{1}{2}\,\text{cbar}^2\,e^{-\text{cbar}} + \frac{1}{3}\,\text{cbar}^3\,e^{-\text{cbar}}$$

Here, the fraction of the network with coverage 3 is counted twice as these regions contribute 2 crossings.

The area of a crossing between fibres with intersection angle ϕ is

$$\frac{\omega^2}{\sin(\phi)}$$

so the expected number of such ϕ-crossings is

```
In[10]:= Simplify[% Sin[φ] / ω²]
```

Out[10]= $\dfrac{\text{cbar}^2\ (3+2\ \text{cbar})\ e^{-\text{cbar}}\ \text{Sin}[\phi]}{6\ \omega^2}$

We noted earlier that the probability density of crossing angles ϕ is $\sin(\phi)$, so the expected number of crossings per unit area is obtained using

In[11]:= **Integrate[% Sin[ϕ], {ϕ, 0, π/2}]**

Out[11]= $\dfrac{\text{cbar}^2\ (3+2\ \text{cbar})\ e^{-\text{cbar}}\ \pi}{24\ \omega^2}$

This differs from Equation 4.15 by a factor

In[12]:= **% / ncrosspua**

Out[12]= $\dfrac{1}{24}\ (3+2\ \text{cbar})\ e^{-\text{cbar}}\ \pi^2$

and this results in a difference in our estimate of about 25 % over the applicable range:

In[13]:= **Plot[%, {cbar, 0, 1}, PlotRange → All]**

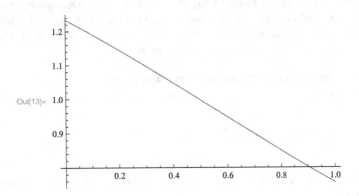

Out[13]=

The different estimates arrived at by the two approaches are due to the first stage of the derivation of Kallmes and Corte [74] computing the probability of crossings between *lines* instead of fibres. Note also that where a fibre crossing occurs close to the end of one or both fibres generating it, the crossing area may be less than $\omega^2 / \sin(\phi)$. More important than the absolute number of crossings is the dependence of this property on the parameters of the network and its constituent fibres, these being essentially the same for both treatments.

4.2.2 Fractional Contact Area

The fractional contact area is defined as the expected fraction of the surface of fibres that contacts other fibres in the network. The measurement often reported in the literature is the analogous relative bonded area (RBA), *i.e.* the expected fraction of the surface of fibres bonded to other fibres in the network, which is important in tensile theories of fibrous materials. Some of the earliest direct measurements of relative bonded area are provided by Page *et al.* [115] who studied paper samples and measured the areas of fibre crossings and their bonded fractions using light microscopy; they reported values of relative bonded area between 0.16 and 0.36. Importantly, Page *et al.* identify also the difficulty in distinguishing between regions of crossing fibres that are bonded and those that are in 'optical contact'; the latter being regions where fibres are sufficiently close that light may pass through them without being scattered so that they appear to be in contact. Since crossing regions may or may not be bonded to each other in a real network, the correct structural parameter for modelling is the fractional contact area and this represents an upper bound on RBA such that it is identical when all areas of contact are bonded.

We derive the fractional contact area of a two-dimensional network following Sampson [143] and consider the contact states of any given fibre within such a two-dimensional network with fibre centres distributed according to a point Poisson process in the plane. Each fibre has an upper and a lower surface and each contact point involves one surface of each contributing fibre. For a two-dimensional fibre network, we assume that fibres make full contact wherever they cross, so the number of surfaces making contact at points with coverage c is

$$n_s(c) = c - 1 . \tag{4.19}$$

The expected fraction of any given fibre in a two-dimensional network that is in contact with other fibres is given by

$$\Phi_{2D} = \frac{1}{\bar{c}} \sum_{c=1}^{\infty} n_s(c)\, P(c) , \tag{4.20}$$

which we implement in *Mathematica* as,

```
In[1]:= P[c_] := PDF[PoissonDistribution[cbar], c]
        Φ2d = Simplify[Sum[(c-1) P[c], {c, 2, ∞}] / cbar]
```

$$\text{Out[2]= } 1 + \frac{-1 + e^{-\text{cbar}}}{\text{cbar}}$$

We note that the same expression was derived by Kallmes *et al.* [76] by considering only regions with coverage up to 3 such that the fraction of the fibre area *not* contacting other fibres is represented by the outer surfaces of regions with coverage 1, 2 and 3, and the fractional contact area is

$$\Phi_{2D} = 1 - \frac{P(1) + P(2) + P(3)}{\bar{c}} . \tag{4.21}$$

From the definition of a two-dimensional network, the fraction of the network with coverage $c > 3$ is negligible, so Kallmes *et al.* [76] assumed

$$P(3) = 1 - (P(0) + P(1) + P(2)) , \tag{4.22}$$

so,

$$\Phi_{2D} = 1 - \frac{1 - P(0)}{\bar{c}} \tag{4.23}$$

$$= 1 - \frac{1 - e^{-\bar{c}}}{\bar{c}} , \tag{4.24}$$

recovering the expression we have derived by accounting for all possible coverages and assuming that all crossings make full contact[1]. Note that the probability of coverage zero is easily measured using image analysis to quantify the fraction of the network not covered by fibres; for a random network this fractional open area corresponds to the probability of pinholes as discussed on page 108 and is given by

$$\epsilon = e^{-\bar{c}} . \tag{4.25}$$

In terms of this parameter, the fractional contact area of a two-dimensional network is therefore:

In[3]:= **cbar = Log[1/ϵ];**
   Apart[Simplify[Φ2d, 0 < ϵ < 1]]

Out[4]= $1 + \dfrac{1 - \epsilon}{\text{Log}[\epsilon]}$

We note also that Equation 4.25 provides also a route to determining the areal density of fibres β_{f}. If the areal density of a network $\bar{\beta}$ has been determined gravimetrically, and the fractional open area has been determined by image analysis, then

$$\beta_{\mathrm{f}} = \frac{\bar{\beta}}{\log(1/\epsilon)} . \tag{4.26}$$

---

[1] In fact, the expression of Kallmes *et al.* can be derived for regions with arbitrary $c \leq n$. Assuming full contact we have $\Phi_{2D} = 1 - \frac{P(1) + P(2) + \ldots + P(n)}{\bar{c}}$; the approximation $P(n) \approx 1 - (P(0) + P(1) + \ldots + P(n-1))$ remains valid and thus, so does Equation 4.24.

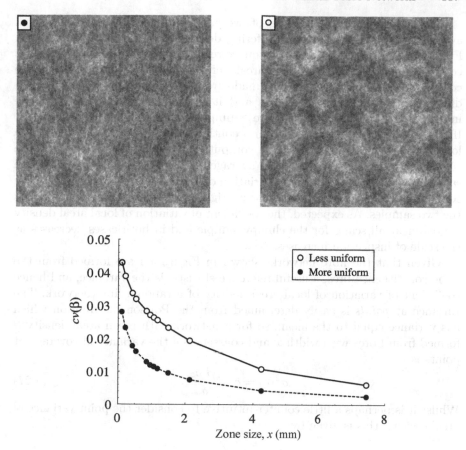

Figure 4.4. Calibrated maps showing in-plane distribution of mass of two paper samples with differing degrees of uniformity. Plot shows the coefficient of variation of local areal density plotted against size of inspection zones

4.2.3 Fractional Between-zones Variance

In real fibrous networks such as paper and many non-woven textiles, the degree of non-uniformity due to clustering can be readily observed by inspecting the network under transmitted light. The texture of a light transmission map of paper closely resembles the distribution of mass within the sheet [45] and we expect this to be the case for other fibrous networks. The distributions of light transmission and mass are not one-to-one however and faithful characterisation of the mass distribution requires calibrated radiography using, for example, β-radiation [78, 79] or soft X-rays [10, 52]. For full discussion of such techniques and experimental conditions, see [24, 163] and for comparisons between different imaging systems, see [4, 80]

Maps of the local areal density of two paper samples with the same global average areal density, but with differing degrees of uniformity are shown in Figure 4.4. Each map represents a square region of side 3 cm and dark regions correspond to high local average areal density, and thus high local average coverage. The samples shown are made from the same fibre type but with different manufacturing conditions and it is immediately evident that the image on the right exhibits a more 'clumpy' texture than that on the left. If the density map is partitioned into contiguous square zones of side x, the local average areal density can be computed for each square zone and the coefficient of variation of the local average areal density determined. As we saw on page 63, the coefficient of variation observed depends on the size of the square inspection zones and the plot in Figure 4.4 shows this dependency for the two samples. As expected, the coefficient of variation of local areal density is greater at all scales for the clumpy sample and in both cases decreases as the scale of inspection increases.

Given that the fibre networks shown in Figure 4.4 are formed from the same constituent fibres, a useful reference statistic is the variance, and hence coefficient of variation of local areal density of a random fibre network. The variance at *points* is easily determined from the Poisson distribution which has variance equal to the mean, so for a network with mean areal density $\bar{\beta}$ formed from fibres with width ω and coarseness δ the variance of coverage at points is

$$\sigma^2(c) = \bar{c} = \frac{\bar{\beta}\,\omega}{\delta}\,. \tag{4.27}$$

Whilst it is perhaps a little counter-intuitive to consider the point variance of areal density, this is given by

$$\sigma^2(\beta) = \sigma^2\left(\frac{c\,\delta}{\omega}\right)$$

$$= \left(\frac{\delta}{\omega}\right)^2 \sigma^2(c)$$

$$= \frac{\delta}{\omega}\,\bar{\beta}\,, \tag{4.28}$$

which is the product of the areal density of a fibre and that of the network. The coefficient of variation of areal density at points is therefore

$$cv(\beta) = \sqrt{\frac{\delta}{\omega\,\bar{\beta}}}\,. \tag{4.29}$$

Real near-planar fibre networks, such as paper and wet-laid non-wovens, typically have a significant structural component in the third dimension but commonly fibre length is significantly greater than network thickness, and for networks formed by filtration and deposition processes, the structures are effectively layered [132]. Accordingly, we may consider the distribution of mass for pure planar processes, since mass and coverage may be considered as projections onto a plane.

As we have seen, when measuring the variability of real structures we cannot sample at points, but must make observations and measurements on finite zones. The variance of local coverage of a random fibre network at a scale of inspection x will always be less than the variance at points of that network by an amount equal to the variance *within* zones. We shall denote the local averages of variables by placing a tilde (˜) over the variable and include the subscript x to denote that these statistics depend upon the scale of inspection. Denoting the variance within zones of side x, $\sigma^2_{w,x}(\tilde{c})$ we have

$$\sigma^2_x(\tilde{c}) = \sigma^2(c) - \sigma^2_{w,x}(\tilde{c}) , \qquad (4.30)$$

which, on rearranging yields

$$\sigma^2_x(\tilde{c}) = \sigma^2(c) \left(1 - \frac{\sigma^2_{w,x}(\tilde{c})}{\sigma^2(c)} \right) . \qquad (4.31)$$

The bracketed term in Equation 4.31 is called the *fractional between-zones variance*, ρ_x; it is the fraction of the point variance that remains as 'between variance' when sampling zones of side x. Thus, ρ_x weights the point variance to yield the variance observed at finite scales of inspection, and we have

$$\sigma^2_x(\tilde{c}) = \sigma^2(c)\,\rho_x \qquad (4.32)$$

$$cv_x(\tilde{c}) = \sqrt{\frac{\rho_x}{\bar{c}}} \qquad (4.33)$$

$$\sigma^2_x(\tilde{\beta}) = \sigma^2(\beta)\,\rho_x = \frac{\delta}{\omega}\,\bar{\beta}\,\rho_x \qquad (4.34)$$

$$cv_x(\tilde{\beta}) = \sqrt{\frac{\delta\,\rho_x}{\omega\,\bar{\beta}}} \qquad (4.35)$$

Before deriving an expression for the fractional between-zones variance, we must introduce the concept of autocorrelation. We have previously encountered correlation as a measure of the association between two random variables. Autocorrelation measures the association between values of a random variable separated by space or time. Qualitatively, we may illustrate the concept by reference to Figure 4.4. If we select a point within one of these textures and then select another a distance r from that point, then for small r we expect the coverage at these points to be similar as a consequence of the underlying texture; as we increase r and move further from our initial point, we increase the likelihood that the coverage of the second point differs significantly from that of the first. More rigorously, the point autocorrelation function $\alpha(r)$ for coverage at points separated by a distance r is given by the ratio of the variance of the number of fibres covering a pair of points separated by a distance r to that of the number of fibres covering one point [27], *i.e.*

$$\alpha(r) = \frac{\sigma^2(c\,|\,r)}{\sigma^2(c\,|\,0)} = \frac{\sigma^2(c\,|\,r)}{\sigma^2(c)}$$

$$= \frac{\sigma^2(c\,|\,r)}{\bar{c}} \tag{4.36}$$

Here we derive expressions for the point autocorrelation function $\alpha(r)$ following Dodson [30, 31]; the treatment provided here draws significantly on the very clear presentation of the theory provided by Schaffnit [148].

Consider a random fibre network with mean coverage $\bar{c}$. If the expected number of fibres per unit area is μ and the fibres have uniform length λ and width ω, then the number of fibres covering a point p is a Poisson variable with mean and variance,

$$\sigma^2(c) = \bar{c} = \mu \lambda \omega \ . \tag{4.37}$$

From Equation 4.36, we seek the variance of coverage of points separated by a distance r, $\sigma^2(c\,|\,r)$. Consider now a point q a distance r from the point p. The number of fibres that cover both p and q is a Poisson variable also with mean and variance $\sigma^2(c\,|\,r) = \bar{c}_{pq}$. To evaluate Equation 4.36 we require an expression for $\bar{c}_{pq}$; $\bar{c}$ being given by Equation 4.37.

The distribution of fibre centres is Poisson with mean μ per unit area which we denote as

$$\mu = \int_{-\frac{\pi}{2}}^{\frac{\pi}{2}} d\mu(\theta) \ , \tag{4.38}$$

where θ is the angle made by the longitudinal axes of fibres to some reference direction. For a uniform distribution of fibre orientations,

$$d\mu(\theta) = \frac{\mu}{\pi} \, d\theta \tag{4.39}$$

Figure 4.5 shows the extreme locations of two fibres with orientation θ covering points p and q separated by a distance r. The line connecting p and q makes an angle ψ with the reference direction. For both p and q to be covered by both fibres, we require that the fibre centres lie within the shaded region. This has area

$$A_{pq} = (\omega - r|\sin(\psi - \theta)|)\,(\lambda - r|\cos(\psi - \theta)|) \ . \tag{4.40}$$

We require also,

$$r|\sin(\psi - \theta)| \le \omega \ \text{ and } \ r|\cos(\psi - \theta)| \le \lambda \tag{4.41}$$

The expected number of fibres that cover both p and q and hence $\sigma^2(c\,|\,r)$ is obtained by integrating Equation 4.40 with respect to θ in the range $-\pi/2 \le \theta \le \pi/2$ satisfying the inequalities given in Equation 4.41. Denoting this set of values Ω, we have

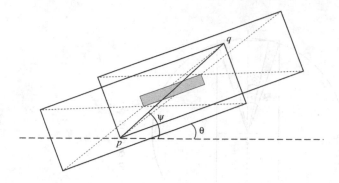

Figure 4.5. Fibres with orientation θ covering two points p and q separated by a distance r. The line connecting p and q has orientation ψ to the reference direction and for both fibres to cover both p and q their centres must lie within the shaded region

$$\sigma^2(c \,|\, r) = \bar{c}_{pq} = \int_\Omega A_{pq} \, d\mu(\theta) . \tag{4.42}$$

It follows that the point autocorrelation function is given by

$$\alpha(r) = \frac{1}{\pi \lambda \omega} \int_\Omega (\omega - r |\sin(\psi - \theta)|) \, (\lambda - r |\cos(\psi - \theta)|) \, d\theta \tag{4.43}$$

Before evaluating this integral in *Mathematica*, we must determine the domain of integration Ω for three cases of interest. To obtain a tractable expression we substitute the variable $\phi = (\psi - \theta)$ and, since both θ and ψ take values between $-\pi/2$ and $\pi/2$, we have

$$\psi - \frac{\pi}{2} \le \phi \le \psi + \frac{\pi}{2}$$

which must satisfy,

$$r |\sin(\phi)| \le \omega$$
$$r |\cos(\phi)| \le \lambda$$

We seek the conditional probability that points p and q, separated by a distance r are both covered by a pair of fibres, *i.e.* the probability that q is covered on the condition that another point p is already covered. Referring to Figure 4.6, we observe that there are three cases to consider; in each case, the shaded region represents that where a fibre centre q must lie in order that it covers the fibre centre p:

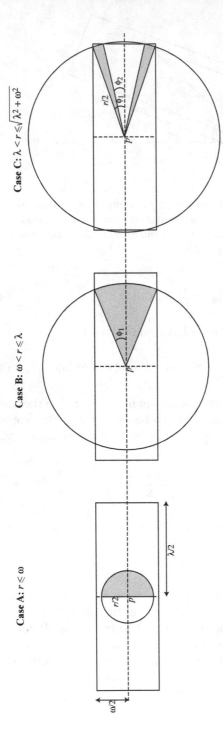

Figure 4.6. Three cases for the domain of integration of the autocorrelation function $\alpha(r)$ in ϕ

Case A: We have $r \leq \omega$, so the domain of integration is

$$-\frac{\pi}{2} \leq \phi \leq \frac{\pi}{2} .$$

Case B: We have $\omega < r \leq \lambda$, so the domain of integration is

$$-\phi_1 \leq \phi \leq \phi_1$$
$$-\arcsin(\omega/r) \leq \phi \leq \arcsin(\omega/r) .$$

Case C: We have $\lambda < r \leq \sqrt{\lambda^2 + \omega^2}$, so the domain of integration is

$$|\phi_2| \leq \phi \leq |\phi_1|$$
$$|\arccos(\lambda/r)| \leq \phi \leq |\arcsin(\omega/r)| .$$

We proceed to use *Mathematica* to evaluate Equation 4.43 using these domains of integration. The first two cases are straightforward:

```
In[1]:= Apq = (ω - r Abs[Sin[φ]]) (λ - r Abs[Cos[φ]]);
        α1 =
        Expand[(1 / (π λ ω)) Integrate[Apq, {φ, -π / 2, π / 2}]]
        α2 = Expand[PowerExpand[(1 / (π λ ω))
            Integrate[Apq, {φ, -ArcSin[ω / r], ArcSin[ω / r]},
            GenerateConditions → False]]]
```

$$\text{Out[2]= } 1 - \frac{2\,r}{\pi\,\lambda} - \frac{2\,r}{\pi\,\omega} + \frac{r^2}{\pi\,\lambda\,\omega}$$

$$\text{Out[3]= } -\frac{2\,r}{\pi\,\omega} - \frac{\omega}{\pi\,\lambda} + \frac{2\,\sqrt{r^2 - \omega^2}}{\pi\,\omega} + \frac{2\,\text{ArcSin}\left[\frac{\omega}{r}\right]}{\pi}$$

For Case C, we take advantage of symmetry and multiply the integral by 2 to account for the two regions of the domain, as illustrated in Figure 4.6:

```
In[4]:= α3 = Expand[PowerExpand[
            (2 / (π λ ω)) Integrate[Apq, {φ, ArcCos[λ / r],
            ArcSin[ω / r]}, GenerateConditions → False]]]
```

$$\text{Out[4]= } \frac{2\,\sqrt{r^2 - \lambda^2}}{\pi\,\lambda} - \frac{r^2}{\pi\,\lambda\,\omega} - \frac{\lambda}{\pi\,\omega} - \frac{\omega}{\pi\,\lambda} +$$

$$\frac{2\,\sqrt{r^2 - \omega^2}}{\pi\,\omega} - \frac{2\,\text{ArcCos}\left[\frac{\lambda}{r}\right]}{\pi} + \frac{2\,\text{ArcSin}\left[\frac{\omega}{r}\right]}{\pi}$$

To simplify subsequent evaluations, we group the three cases into a piece-wise function, bearing in mind that for unstated conditions, the function evaluates to zero.

In[5]:= $\alpha[\lambda\_,\ \omega\_,\ r\_]\ :=$

$$\text{Piecewise}\Big[\Big\{\Big\{1-\frac{2\,r}{\pi\,\lambda}-\frac{2\,r}{\pi\,\omega}+\frac{r^2}{\pi\,\lambda\,\omega},\ 0\le r\le\omega\Big\},$$

$$\Big\{-\frac{2\,r}{\pi\,\omega}-\frac{\omega}{\pi\,\lambda}+\frac{2\,\sqrt{r^2-\omega^2}}{\pi\,\omega}+\frac{2\,\text{ArcSin}\big[\frac{\omega}{r}\big]}{\pi},\ \omega<r\le\lambda\Big\},$$

$$\Big\{\frac{2\,\sqrt{r^2-\lambda^2}}{\pi\,\lambda}-\frac{r^2}{\pi\,\lambda\,\omega}-\frac{\lambda}{\pi\,\omega}-\frac{\omega}{\pi\,\lambda}+\frac{2\,\sqrt{r^2-\omega^2}}{\pi\,\omega}-$$

$$\frac{2\,\text{ArcCos}\big[\frac{\lambda}{r}\big]}{\pi}+\frac{2\,\text{ArcSin}\big[\frac{\omega}{r}\big]}{\pi},\ \lambda<r\le\sqrt{\lambda^2+\omega^2}\ \Big\}\Big\}\Big]$$

A plot of $\alpha(r)$ against the separation of points r in units of the fibre length λ shows the autocorrelation function to decay with r. The example shown here is for fibres of aspect ratio 50; for larger aspect ratios, the decay is steeper.

In[6]:= $A = 50;\ \lambda = 1;\ \omega = \lambda\,/\,A;$
$\text{Plot}[\alpha[\lambda,\ \omega,\ r],\ \{r,\ 0,\ .4\},$
$\quad\text{PlotRange} \to \text{All},\ \text{AxesLabel} \to \{"r",\ "\alpha(r)"\}]$
$\text{Clear}[\lambda,\ \omega]$

Out[7]=

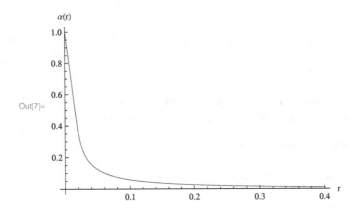

We state then that the point autocorrelation function $\alpha(r)$ for a random network of fibres with length λ and width ω is given by

$$\alpha(r) = \begin{cases} 1 - \frac{2}{\pi}\left(\frac{r}{\lambda} + \frac{r}{\omega} - \frac{r^2}{2\omega\lambda}\right) & 0 < r \leq \omega \\[2mm] \frac{2}{\pi}\left(\arcsin\left(\frac{\omega}{r}\right) - \frac{\omega}{2\lambda} - \frac{r}{\omega} + \sqrt{\frac{r^2}{\omega^2} - 1}\right) & \omega < r \leq \lambda \\[2mm] \frac{2}{\pi}\left(\arcsin\left(\frac{\omega}{r}\right) - \arccos\left(\frac{\lambda}{r}\right) - \frac{\omega}{2\lambda}\right. & \\[2mm] \left. -\frac{\lambda}{2\omega} - \frac{r^2}{2\lambda\omega} + \sqrt{\frac{r^2}{\lambda^2} - 1} + \sqrt{\frac{r^2}{\omega^2} - 1}\right) & \lambda < r \leq \sqrt{\lambda^2 + \omega^2} \\[2mm] 0 & \text{otherwise.} \end{cases}$$

$$(4.44)$$

Having obtained the point autocorrelation function, we return to the fractional between-zones variance, ρ_x, introduced on page 119. Dodson [30, 31] showed that the fractional between-zones variance, ρ_x could be obtained by integrating the point autocorrelation function $\alpha(r)$ for all r. Denoting the probability density of r, $b(r)$ we have therefore

$$\rho_x = \int_\Omega \alpha(r)\, b(r)\, \mathrm{d}r \ . \tag{4.45}$$

In Section 3.2.2 we derived the probability density function for separation of pairs of points occurring at random in a square region. This probability density, as given by Equation 3.10, can be used to describe the distribution of distances between fibre centres:

```
In[9]:= b[x_, r_] := Piecewise[{{2 r (π x² - 4 x r + r²) / x⁴,  0 < r ≤ x},
        {2 r (4 x √(r² - x²) - r² - x² (2 + π - 4 ArcSin[x / r])) / x⁴,
        x < r ≤ √2 x}}]
```

This provides also the domain of integration required to compute the fractional between-zones variance, so we have

$$\rho_x = \int_0^{\sqrt{2}\,x} \alpha(r)\, b(r)\, \mathrm{d}r \ . \tag{4.46}$$

For certain conditions, the partial integral can be determined analytically, for example if $0 < r < \omega < x$:

```
In[11]:= Expand[Integrate[PiecewiseExpand[α[λ, ω, r] b[x, r]],
        {r, 0, ω}, Assumptions → 0 ≤ r < ω < x]]
```

$$\text{Out[11]}= \ -\frac{4\,\omega^2}{3\,x^2} + \frac{\pi\,\omega^2}{x^2} - \frac{8\,\omega^3}{3\,x^3} + \frac{4\,\omega^3}{\pi\,x^3} -$$

$$\frac{5\,\omega^3}{6\,x^2\,\lambda} + \frac{\omega^4}{2\,x^4} - \frac{4\,\omega^4}{5\,\pi\,x^4} + \frac{12\,\omega^4}{5\,\pi\,x^3\,\lambda} - \frac{7\,\omega^5}{15\,\pi\,x^4\,\lambda}$$

Typically however, for integration over the full domain, numerical methods are required. It is convenient to define a function with arguments given by the variables required for evaluation:

```
In[12]:= ρx[λ_, ω_, x_, r_] :=
        NIntegrate[α[λ, ω, r] b[x, r], {r, 0, √2 x}]
```

Each of the variables in this function has dimensions of length, so the values input for evaluation must be given in the same units. For example, a typical staple used in a non-woven fabric may have length 1 cm and width 20 μm. The fractional between-zones variance observed at a 1-mm scale of inspection is therefore calculated using,

```
In[13]:= ρx[10, .02, 1, r]
```

```
Out[13]= 0.0182827
```

and from Equation 4.32 we may state that the variance of local coverage of a random network of such fibres at the 1-mm scale of inspection is just under 2 % of the variance of coverage at points.

To investigate the influence of each variable on the fractional between-zones variance, we evaluate the integral within a **Table** command. In the following example we compute the influence of inspection zone size on ρ_x for fibres of length 1 mm and width 20 μm, which may be considered typical of a papermaking fibre. To reduce the number of integrations, the first line of code specifies the x with larger intervals as the function becomes less sensitive to x; note also the use of **Prepend** to include the known analytic result that $\rho_x = 1$ when $x = 0$.

```
In[14]:= xtab = Join[Range[.05, .5, .05],
        Range[.6, 1, .1], Range[1, 5, 1]];
    xρtab = Prepend[Table[{xtab[[i]], ρx[1, .02,
        xtab[[i]], r]}, {i, 1, Length[xtab]}], {0, 1}];
    ListLinePlot[xρtab, PlotRange → All,
    AxesLabel → {"x", "ρx"}]
```

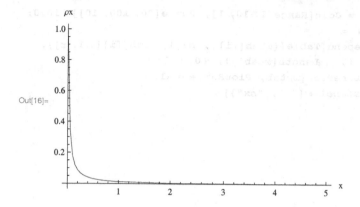

As expected, the variance decreases with increasing inspection zone size and importantly, plots on logarithmic axes do not yield a linear dependence. We obtain the dependence of ρ_x on fibre length in a similar way for fibres of width 20 µm at the 1-mm scale of inspection:

```
In[17]:= λtab = Join[Range[.05, .5, .05],
          Range[.6, 2, .1], Range[3, 10, 1]];
        λρtab = Prepend[Table[{λtab[[i]], ρx[λtab[[i]],
          .02, 1, r]}, {i, 1, Length[λtab]}], {0, 0}];
        ListLinePlot[λρtab, PlotRange → All,
          AxesLabel → {"λ", "ρx"}]
```

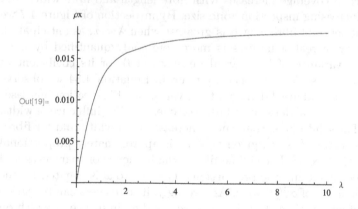

Finally, we observe an approximately linear dependence of ρ_x on fibre width:

```
In[20]:= ωtab = Join[Range[1, 10, 1], Range[20, 100, 10]] / 1000;
         ωρtab =
            Prepend[Table[{ωtab[[i]], ρx[1, ωtab[[i]], 1, r]},
               {i, 1, Length[ωtab]}], {0, 0}];
         ListLinePlot[ωρtab, PlotRange → All,
            AxesLabel → {"ω", "ρx"}]
```

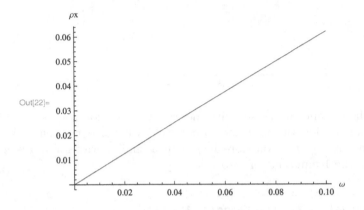

Collected results for the influence of the variables affecting the fractional between-zones variance are shown in Figure 4.7. From Equation 4.32, we observe that the variance of coverage is given by the product of the mean coverage and the fractional between-zones variance. Accordingly, we may state that the variance of coverage increases with fibre length and fibre width and decreases with increasing inspection zone size. By inspection of Figure 4.7 we note that the influence of fibre length is greatest when $\lambda < x$, as anticipated.

The uniformity of real structures is more commonly quantified by measurements of the variance of local areal density $\sigma_x^2(\tilde{\beta})$, or its coefficient of variation $cv_x(\tilde{\beta})$, at a scale of inspection x. From Equation 4.34 we observe that the variance of local areal density at a given scale of inspection increases with increasing mean areal density and fibre coarseness. The influence of width is more subtle. For solid fibres, coarseness increases with width and for fibres of circular cross section $\delta \propto \omega^2$; given that ρ_x is approximately proportional to ω also, then $\sigma_x^2(\tilde{\beta}) \propto \omega^2$. For hollow fibres, the influence of coarseness and width can be decoupled and the proportionality of ρ_x to ω is negated by the inverse proportionality of $\sigma_x^2(\tilde{\beta})$ to ω. Accordingly, if coarseness can be varied independently of width, then there is no corresponding influence of width on the variance of local areal density. The widely observed tendency of networks formed from wide fibres to be non-uniform can be identified therefore as being as a consequence of the influence of width on coarseness. The dependencies we have identified carry through also to the coefficient of variation of local areal density, as given by Equation 4.35; this measure of uniformity decreases with increasing mean areal density and increases with length and coarseness. The effects of length, coarseness and mean areal density on the uniformity

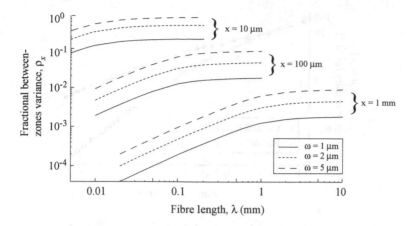

Figure 4.7. Collected results for the influence of fibre length and width and inspection zone size on the fractional between-zones variance, ρ_x

of a random fibre network are attributable to a common influence: changing these variables directly influences the number of fibre centres per unit area.

We define the following functions to compute the variance and coefficient of variation of local areal density using Equations 4.34 and 4.35, respectively:

```
In[23]:=  Varβ[λ_, ω_, x_, βbar_, δ_] :=
             δ βbar ρx[λ, ω, x, r] / (ω / 1000)
          CVβ[λ_, ω_, x_, βbar_, δ_] :=
             √‾δ ρx[λ, ω, x, r] / (ω βbar / 1000)
```

As previously, we require the same units for fibre length, fibre width and inspection zone size. Mean areal density is typically reported with units $g\,m^{-2}$; accordingly, for correct evaluation of the above code we specify fibre length, fibre width and inspection zone size in mm, mean areal density in $g\,m^{-2}$ and coarseness with units of $g\,m^{-1}$. At the 1-mm scale of inspection, the variance and coefficient of variation of local areal density of a network with mean areal density $60\ g\,m^{-2}$, formed from fibres of length 4 mm, width 10 μm and coarseness $1.5 \times 10^{-4}\,g\,m^{-1}$ is therefore computed as,

```
In[25]:=  Varβ[4, .01, 1, 60, 1.5 10⁻⁴]
          CVβ[4, .01, 1, 60, 1.5 10⁻⁴]
```

Out[25]= 7.80033

Out[26]= 0.0465484

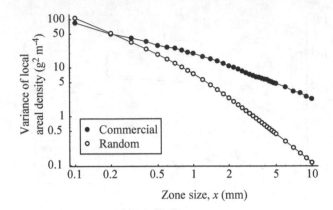

Figure 4.8. Variance of local areal density plotted against zone size for a commercial sample and a random network formed from the same fibres at the same mean areal density

so we have $\sigma_x^2(\tilde{\beta}) = 7.8\,\mathrm{g^2\,m^{-4}}$ and $cv_x(\tilde{\beta}) = 4.65\,\%$.

Where there is a distribution of lengths, $\rho_{x,i}$ should be determined for each fibre length class λ_i and, as the variance contributions of each length fraction are independent, they are additive. So the total fractional between-zones variance is given by

$$\rho_x = \sum_i^n m_i\,\rho_{x,i}\;,\tag{4.47}$$

where m_i is the mass fraction of fibres with class centre λ_i.

Figure 4.8 shows the variance of local areal density of an industrially manufactured paper sample, as measured by contact β-radiography and image analysis, plotted against zone size on logarithmic scales; the data for the commercial sample are represented by solid markers. On the same axes, the variance of local areal density of a random network of the same constituent fibres is plotted using open markers. Our first observation is that the variance of local areal density of the commercial sample is greater than that of the random network at all scales above about 0.2 mm; this is typical of almost all commercially formed papers [27] and has been observed for non-woven filters also [165].

Commercial papermaking involves handling fibrous slurries of sufficiently high concentration that fibres are in continuous mutual contact such that they cannot be deposited independently of each other, thus breaking one of our conditions for randomness. The higher variance observed in Figure 4.8 is a consequence of these fibre interactions during the manufacturing process resulting in a more clumpy structure. A useful quantifier of such departures from randomness is a *variance ratio* obtained by dividing the observed variance by that calculated for a random fibre network with the same mean areal

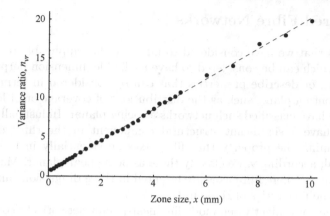

Figure 4.9. Variance ratio plotted against zone size for the commercial sample shown in Figure 4.8

density, formed from the same constituent fibres:

$$\text{Variance ratio, } n_{vr}(x) = \frac{\sigma_x^2(\tilde{\beta})_{\text{measured}}}{\sigma_x^2(\tilde{\beta})_{\text{random}}} \qquad (4.48)$$

So, at a given zone size, x we have

$$n_{vr} = 1 \qquad \text{Same uniformity as random}$$
$$n_{vr} > 1 \qquad \text{Less uniform (more clustered) than random}$$
$$n_{vr} < 1 \qquad \text{More uniform (less clustered) than random}$$

Figure 4.9 shows a plot of the variance ratio against zone size using the data shown in Figure 4.8. Again, the behaviour shown here is typical for commercially manufactured paper, though some non-linearity may be observed at scales of inspection above about 4 mm [27]. The dependence of the variance ratio on the scale of inspection is consistent with a cooperative process during forming resulting in the resultant structure being closer to that of a random network at smaller scales. Such cooperative effects have been identified experimentally in paper-forming processes, with fibres being preferentially deposited in sparse regions as the structure evolves [62, 95, 111, 145]. We expect the same effects to be present during any filtration type network-forming process such as wet-laying or air-laying of non-wovens. In electrospinning, such effects are unlikely since fibres are produced at the same stage as the network. Indeed, given the inherent electrostatic charge associated with such processes, it is possible that fibre repulsions result in networks having a variance ratio less than 1; this conjecture remains to be tested, though we note the observation of Matthews *et al.* [101] that the uniformity of electrospun networks of collagen fibres could be influenced by process variables. We shall consider models for departures from randomness, as identifiable through changes in the variance ratio, in more detail in Chapter 6.

4.3 Layered Fibre Networks

The models that we have considered so far are either applicable to very thin networks, which can be considered to have negligible dimension perpendicular to the plane, or describe properties that can be considered in terms of their projection onto a plane, such as the distributions of coverage or of local areal density. We have classified such networks as being planar. Industrially realised structures have a significant structural component in the third dimension whilst retaining the property that fibre axes lie essentially in the plane of the material; accordingly, we classify these as being near-planar. Materials do exist, of course, where fibre axes are oriented in three dimensions and we will consider some properties of these in Chapter 7.

It is often convenient when modelling near-planar networks to consider the properties of structures resulting from superimposing several planar networks. Inevitably, the underlying assumption for such multi-planar models is that the materials we seek to model exhibit structures which are effectively layered. Qualitatively, most of us have observed that it is often easy to remove layers of fibre from paper using adhesive tapes; similarly many non-woven textiles may be readily split into layers once a split has been initiated by picking at an edge. Careful experimental studies have been presented for paper showing that this material tends to exhibit a layered structure and not a felted one, where fibres might extend from one surface of the web to the other [63, 132]. Accordingly, multi-planar models for paper structures are widely reported [2, 59, 75, 156]. Since many non-woven textiles are manufactured using wet- and air-laying processes similar to papermaking, these are often considered to be layered for modelling purposes also [166, 169, 170].

There is some evidence that webs formed from high-concentration fibre suspensions exhibit some fibre entanglement and are thus more felted than layered [61, 110]. When forming such structures, three-dimensional fibre networks are effectively collapsed such that fibres lie essentially parallel to the plane of the network and layered models can be considered applicable to all but the most felted structures. Note also that in many cases, our models will consider the statistics of networks at points and so the extent of layering or felting observed at finite scales is not relevant as long as fibre axes lie essentially in the plane of the network at any given point. In the following sections we expand many of the models derived in Chapter 3 to extend their applicability to near-planar networks; we develop also expressions for structural characteristics of this class of materials that are not evident in planar structures.

4.3.1 Fractional Contact Area

In Section 4.2.2 we showed that the fractional contact area of a planar fibre network with mean coverage $\bar{c}$ is given by

$$\Phi_{2D} = 1 - \frac{1 - e^{-\bar{c}}}{\bar{c}} , \qquad (4.49)$$

and this can be expressed in terms of the fractional open area of the network, $\epsilon = e^{-\bar{c}}$:

$$\Phi_{2D} = 1 + \frac{1 - \epsilon}{\log(\epsilon)} . \qquad (4.50)$$

We will consider such two-dimensional networks to form the constituent layers of a multi-planar structure. The number of layers required to model a network of mean coverage $\bar{c}$ is

$$n = \frac{\bar{c}}{\bar{c}_{2D}} , \qquad (4.51)$$

where $\bar{c}_{2D}$ is the coverage of a two-dimensional network constituting a layer.

Classical approaches to multi-planar modelling had limited applicability since the mean coverage of a layer $\bar{c}_{2D}$ was typically defined arbitrarily, either in terms of the areal density of fibres, β_f [75] or in terms of limiting coverage of a two-dimensional network [2, 59]. The relationship between fractional open area and mean coverage however, provides a definition of the mean coverage of a layer in terms of the network porosity. Since the porosity of the network is the three-dimensional analogue of the fractional open area, we denote it also by ϵ and the coverage of a layer is therefore $\bar{c}_{2D} = \log(1/\epsilon)$, so the number of layers required to model a network with mean coverage $\bar{c}$ is

$$n = \frac{\bar{c}}{\log(1/\epsilon)} . \qquad (4.52)$$

Note that whilst the fractional open area of high-coverage networks may still be computed as $\epsilon = e^{-\bar{c}}$ this does not correspond to the network porosity, since networks of arbitrary realisable coverage may exhibit any porosity and these two parameters are coupled for two-dimensional networks only.

Consider now a structure formed by superimposing two-dimensional networks with independent and identical distributions of contacts such that they have the same fractional contact area, Φ_{2D}. Each layer of such a network is assumed to make new contacts with the layers immediately above and below it only. Although we may conceive of fibres penetrating through more than one layer to contact fibres in another through, *e.g.* consolidation processes such as pressing or calendering, our definition of a layer renders this unnecessary since such penetration reduces porosity and this is the independent variable in our model.

We consider first three layers within a multi-planar structure; the treatment is guided by the approach of Sampson [143], but here it has the advantage that the resultant expressions are simpler and applicable over the full range of network porosities. A given fibre in the central layer may form additional contacts with the surrounding layers in those regions where its own layer has no contact on either side, or is in contact with other fibres above or below it only. The fraction of the projected solid area making contacts between layers is $(1 - \epsilon)^2$ and the fraction of the fibre surface that is available for additional contacts is $(1 - \Phi_{2D})$. Accordingly, for a network with infinite coverage formed from the superposition of an infinite number two-dimensional of layers, we have,

$$\Phi_\infty = \Phi_{2D} + (1 - \varepsilon)^2 (1 - \Phi_{2D}) \ . \tag{4.53}$$

Inputting Equations 4.50 and 4.53 to *Mathematica* yields Φ_∞ in terms of porosity only:

```
In[1]:= Φ2d = 1 + (1 - ε) / Log[ε];
        Φinf = Φ2d + (1 - ε)² (1 - Φ2d);
        Φinf = FullSimplify[Φinf, 0 < ε < 1]
```

$$Out[3]= \ 1 + \frac{\epsilon \ (2 + (-3 + \epsilon) \ \epsilon)}{Log \ [\epsilon]}$$

A plot of Φ_∞ against porosity shows a non-linear decrease of fractional contact area with increasing porosity and when

$$\epsilon = 0 \qquad \Phi_\infty = 1$$
$$\epsilon = 1 \qquad \Phi_\infty = 0$$

as expected.

```
In[4]:= Plot[Φinf, {ε, 0.0001, 1}, AxesLabel → {"ε", "Φ∞"}]
```

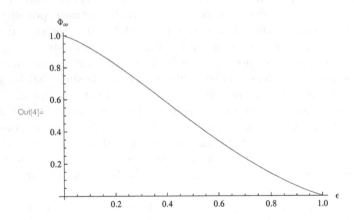

When we consider multi-planar networks of finite thickness we must take account of the fraction of the total fibre length which is located at the surfaces of the network and that can be in contact with other fibres on one side only. At points with coverage c, the fraction of fibre surfaces available for contact with other fibres is $(c - 1)/c$. Denoting the Poisson probability of coverage c

as $P(c)$, the fraction of the network covered by fibres is $(1 - P(0))$ so the fraction of the network as a whole that is available for contact is

$$f = \frac{1}{1 - P(0)} \sum_{c=2}^{\infty} \frac{c-1}{c} P(c) , \qquad (4.54)$$

which we evaluate as:

```
In[5]:= P[c_] := PDF[PoissonDistribution[cbar], c]
        f = FullSimplify[
            1 / (1 - P[0]) Sum[(c - 1) / c P[c], {c, 2, ∞}], cbar > 0]
```

$$\text{Out[6]}= 1 + \frac{\text{EulerGamma} - \text{ExpIntegralEi}[\text{cbar}] + \text{Log}[\text{cbar}]}{-1 + e^{\text{cbar}}}$$

where **EulerGamma** is Euler's constant (≈ 0.577) and **ExpIntegralEi** is the exponential integral function.

As expected, the fraction of the network that is available for contact, f, increases towards an asymptote at $f = 1$ with increasing mean coverage $\bar{c}$. Note also that for $\bar{c}$ greater than about 10, we have the expected approximation, $f \approx 1 - \frac{1}{\bar{c}}$.

```
In[7]:= LogLinearPlot[{f, 1 - 1 / cbar, 1},
        {cbar, 0.01, 100}, PlotStyle → {{}, Dashed, Dashed},
        PlotRange → {0, 1}, AxesLabel → { "c̄", "f"}]
```

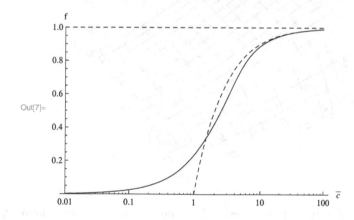

The fractional contact area of a network of finite mean coverage $\bar{c}$ is given by

$$\Phi_c = f \, \Phi_\infty , \qquad (4.55)$$

such that $0 \le \Phi_c \le f$.

In[8]:= **Φc = f Φinf**

Out[8]= $\left(1 + \dfrac{\text{EulerGamma} - \text{ExpIntegralEi}[\text{cbar}] + \text{Log}[\text{cbar}]}{-1 + e^{\text{cbar}}}\right)$

$\left(1 + \dfrac{\epsilon\,(2 + (-3 + \epsilon)\,\epsilon)}{\text{Log}[\epsilon]}\right)$

The approximate result should be sufficient for networks with mean coverage $\bar{c}$ greater than about 10. So the fractional contact area of a network with porosity ϵ can be approximated by

$$\Phi_c \approx \left(1 - \frac{1}{\bar{c}}\right)\left(1 + \frac{\epsilon\,(2 - (3 - \epsilon)\,\epsilon)}{\log(\epsilon)}\right). \qquad (4.56)$$

We render a surface showing the dependence of Φ_c on mean coverage and porosity using **Plot3D**:

In[9]:= **Plot3D[Φc, {ε, 0, 1}, {cbar, 0, 20},**
AxesLabel → {"ε", "c̄", "Φc"}]

Out[9]=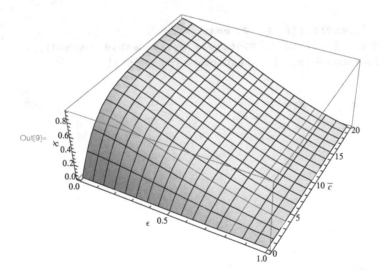

First inspection of this surface reveals that the fractional contact area Φ_c is most sensitive at low coverages. To find the region of greatest sensitivity to porosity, we must differentiate Φ_∞ with respect to ϵ and plot the derivative:

In[10]:= **Plot[Evaluate[D[Φinf, ε]], {ε, 0, 1}]**

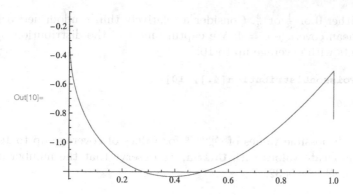

The location of the steepest point in the gradient is found from the second derivative:

In[11]:= **FindRoot[D[Φinf, {ε, 2}] == 0, {ε, 0.4}]**

Out[11]= $\{\epsilon \rightarrow 0.385681\}$

So we observe that fractional contact area is most sensitive to porosity in the range $0.2 \leq \epsilon \leq 0.6$. We note that for tissue engineering applications, the fraction of the fibre surface that is *not* in contact with other fibres, and that is thus available for surface modification and for cells to bind to, is of interest [96, 97, 104]; this fraction is, of course, $1 - \Phi_c$ and thus exhibits the same dependencies on porosity and coverage.

4.3.2 In-plane Distribution of Fractional Contact Area

Above any point in the plane of support of a near-planar fibre network, inter-faces exist between fibres at which contacts may exist with probability Φ_∞, the alternative being the presence of an inter-fibre void with probability $(1-\Phi_\infty)$. The number of interfaces between fibres at a point with coverage c is

$$n_{\text{int}} = c - 1 . \tag{4.57}$$

So the expected number of contacts at points with coverage c is

$$n_{\text{con}} = n_{\text{int}} \, \Phi_\infty = (c-1) \, \Phi_\infty , \tag{4.58}$$

such that the expected fractional contact area at a point with coverage c is

$$\bar{\Phi}_c^{\text{points}} = \frac{c-1}{c} \, \Phi_\infty , \tag{4.59}$$

The fractional contact at points is limited to discrete values however. So, at a point with coverage $c = 2$ we have just one interface and the fractional contact area Φ_c^{points} is either 0 or $\frac{1}{2}$; similarly if $c = 3$ then we have two interfaces and Φ_c^{points} is either 0 or $\frac{1}{3}$ or $\frac{2}{3}$. Consider a relatively thin random network of fibres with mean coverage $\bar{c} = 4$; We capture most of the distribution by considering points with coverage up to 10:

In[1]:= `CDF[PoissonDistribution[4.], 10]`

Out[1]= 0.99716

Considering all possible values of Φ_c^{points} for values of coverage up to 10, and removing duplicate values with **Union**, we observe that the number of possible values of Φ_c^{points} is 32:

In[2]:= ```
Union[
    Flatten[Table[nint / c, {c, 1, 10}, {nint, 0, c - 1}]]]
Length[%]
```

Out[2]= $\left\{ 0, \dfrac{1}{10}, \dfrac{1}{9}, \dfrac{1}{8}, \dfrac{1}{7}, \dfrac{1}{6}, \dfrac{1}{5}, \dfrac{2}{9}, \dfrac{1}{4}, \dfrac{2}{7}, \dfrac{3}{10}, \dfrac{1}{3}, \dfrac{3}{8}, \dfrac{2}{5}, \dfrac{3}{7}, \dfrac{4}{9}, \right.$
$\left. \dfrac{1}{2}, \dfrac{5}{9}, \dfrac{4}{7}, \dfrac{3}{5}, \dfrac{5}{8}, \dfrac{2}{3}, \dfrac{7}{10}, \dfrac{5}{7}, \dfrac{3}{4}, \dfrac{7}{9}, \dfrac{4}{5}, \dfrac{5}{6}, \dfrac{6}{7}, \dfrac{7}{8}, \dfrac{8}{9}, \dfrac{9}{10} \right\}$

Out[3]= 32

The following code computes a coverage, $c_{\text{max}}$ for a range of $\bar{c} \leq 50$ such that the fraction of the network with coverage greater than $c_{\text{max}}$ is of order 1 %. These values are then used to compute the number of possible values of $\Phi_c^{\text{points}}$ expected in a network with mean coverage $\bar{c}$. We observe an exponential increase in the number of possible values of $\Phi_c^{\text{points}}$ when plotted against mean coverage:

In[4]:= ```
cmax = Round[Table[c /. FindRoot[
        Simplify[CDF[PoissonDistribution[cbar], c], c ∈
            Integers] == .99, {c, cbar}], {cbar, 1, 50}]];
ListPlot[Table[{i, Length[Union[Flatten[Table[nint / c,
        {c, 1, cmax[[i]]}, {nint, 0, c - 1}]]]]},
    {i, 1, 50}], AxesLabel → {"c̄",
    "Number of possibilities"}]
```

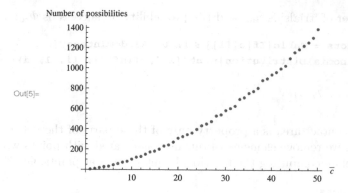

Importantly, the discrete random variable representing the fractional contact area at points does not have a uniform interval between possible values, complicating any analytic approach to determining its probability function. Accordingly, we proceed using a Monte Carlo method. As with earlier examples, first we develop our code by considering only a small number of points, facilitating easy inspection of the outputs generated. Our independent variables are the mean coverage of the network $\bar{c}$ and its porosity ϵ which influences the probability of contact between vertically adjacent fibres, Φ_∞, as given on page 134:

```
In[1]:= n = 5;
        cbar = 5;
        ε = 0.5;
        Φinf = 1 + (ε (2 + (-3 + ε) ε)) / Log[ε];
```

We generate a list of coverages at points c drawn from the Poisson distribution with mean $\bar{c}$ and from this a list of the number of interfaces between vertically adjacent fibres n_{int}, bearing in mind that where coverage is 0 or 1, there are no such interfaces.

```
In[5]:= SeedRandom[5]
        c = RandomInteger[PoissonDistribution[cbar], n]
        nint = Table[If[c[[i]] ≤ 1, 0, c[[i]] - 1], {i, 1, n}]
```

```
Out[6]= {0, 2, 11, 10, 3}
```

```
Out[7]= {0, 1, 10, 9, 2}
```

If we assume that the porosity at points is independent of their coverage, then we may assume that each interface between fibres has equal likelihood of making contact with the fibres generating it. We introduced the concept of Bernoulli trials on page 31 and we use this to generate a list of the number of interfaces that make contact, such that contacts have a binomial distribution where the number of 'trials' is n_{int} and the probability of 'success' is Φ_∞:

```
In[8]:= ncontacts = Table[If[c[[i]] ≤ 1, 0, RandomInteger[
          BinomialDistribution[nint[[i]], Φinf]]], {i, 1, n}]
```

```
Out[8]= {0, 1, 2, 7, 0}
```

Since fractional contact area is a property only of those parts of the network covered by fibre, we remove elements of our lists associated with points with coverage zero before computing the fractional contact area at points, Φ_c^{points}:

```
In[9]:= cncontacts =
          Cases[Transpose[{c, ncontacts}], {x_, y_} /; x ≠ 0]
        cncontacts = Transpose[cncontacts];
        c = cncontacts[[1]];
        ncontacts = cncontacts[[2]];
        Φcpoints = ncontacts / c
```

```
Out[9]= {{2, 1}, {11, 2}, {10, 7}, {3, 0}}
```

$$Out[13]= \left\{ \frac{1}{2}, \frac{2}{11}, \frac{7}{10}, 0 \right\}$$

We collect our code to define a function **FCAMonteCarlo** permitting its execution for the range of independent variables $\bar{c}$ and ϵ; in the example given here, the number of points considered has been set to 50,000 which seems to be sufficient to give stable statistics:

```
In[14]:= Clear[ε, cbar, n, Φinf]
         FCAMonteCarlo[cbar_, ε_] :=
           (n = 50 000; Φinf = 1 + (ε (2 + (-3 + ε) ε)) / Log[ε];
         SeedRandom[5];
           c = RandomInteger[PoissonDistribution[cbar], n];
           nint = Table[If[c[[i]] ≤ 1, 0, c[[i]] - 1], {i, 1, n}];
         ncontacts = Table[If[c[[i]] ≤ 1,
                 0, RandomInteger[BinomialDistribution[
                   nint[[i]], Φinf]]], {i, 1, n}];
         cncontacts = Cases[Transpose[{c, ncontacts}],
                 {x_, y_} /; x ≠ 0];
           cncontacts = Transpose[cncontacts];
           c = cncontacts[[1]];
           ncontacts = cncontacts[[2]];
           Φcpoints = ncontacts / c;)
```

So to calculate the fractional contact area of 50,000 points within a network
with mean coverage $\bar{c} = 5$ and porosity $\epsilon = 0.5$ we evaluate

```
In[16]:= FCAMonteCarlo[5, .5]
```

This yields no output, but has generated a list representing the fractional
contact area at 50,000 points in the network from which we may calculate the
mean, variance, *etc.* The expected fractional contact area $\bar{\Phi}_c$ is less than Φ_∞
by a factor given by the fraction of the network as a whole that is available
for contact f, as derived on page 134:

```
In[17]:= cbar = 5;
        Φcbar = Mean[N[Φcpoints]]
        Φinf
        f = (1 + (EulerGamma - ExpIntegralEi[cbar] + Log[cbar]) /
               (e^cbar - 1));
        Φcbar / f
        cbar =.
```

```
Out[18]= 0.338997
```

```
Out[19]= 0.458989
```

```
Out[21]= 0.456728
```

To generate a bar chart of the fractional contact area at points we must use
GeneralizedBarChart to handle our discrete data with non-uniform intervals. Such a chart reveals the highly discontinuous nature of the distribution
of inter fibre contact in the plane:

```
In[23]:= Needs["BarCharts`"]
        counts = Sort[Tally[Φcpoints]];
        bins = Transpose[counts][[1]];
        barwidths =
          Table[Evaluate[Min[Table[bins[[i + 1]] - bins[[i]],
              {i, 1, Length[bins] - 1}]]], {Length[bins]}];
        frequencies = N[Transpose[counts][[2]] / n];
        GeneralizedBarChart[
         Transpose[{bins, frequencies, barwidths}],
         Frame → {True, True, False, False},
         FrameLabel → {"Φc", "Frequency"},
         PlotRange → {{0, 1}, All}]
```

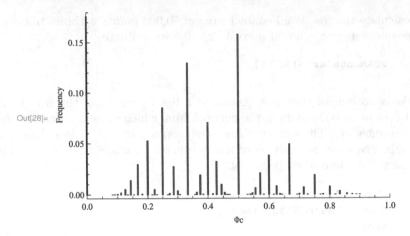

We obtain the mean, variance, coefficient of variation and skewness of our distribution in the usual way:

In[29]:= **Mean[N[Φcpoints]]**
       Variance[N[Φcpoints]]
       Sqrt[%] / %%
       Skewness[N[Φcpoints]]

Out[29]= 0.338997

Out[30]= 0.0464649

Out[31]= 0.635868

Out[32]= -0.0552078

Collected results showing influence of mean coverage and porosity on the distribution of fractional contact area are shown in Figure 4.10. At low mean coverages, we observe a negative skew at low porosities which correspond to a high fractional contact area and a positive skew at higher porosities; at high mean coverages, the distribution is more symmetrical. Note also the high frequency of $\Phi_c^{\mathrm{points}} = \frac{1}{2}$ which is permissible at points with odd point coverage c such that $n_{\mathrm{int}} = (c-1)$ is even.

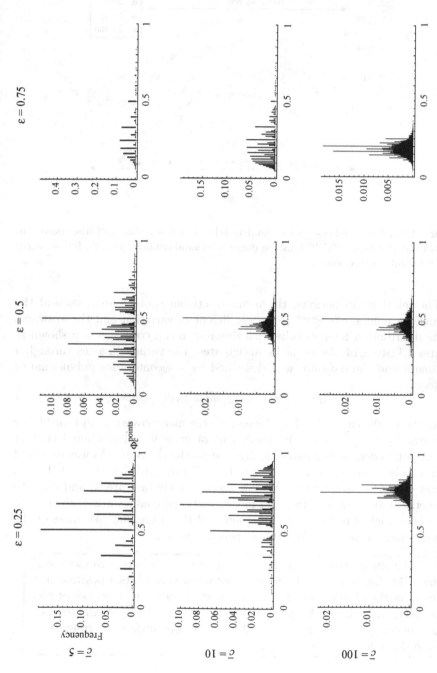

Figure 4.10. Influence of mean network coverage and porosity on the distribution of fractional contact area at points

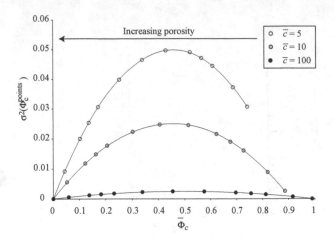

Figure 4.11. Collected results for the dependence of the variance of fractional contact area at points, $\sigma^2(\Phi_c^{\text{points}})$ on the mean fractional contact area, $\bar{\Phi}_c$ for networks with different mean coverage.

The relationship between the mean fractional contact area, $\bar{\Phi}_c$ and the variance at points, $\sigma^2(\Phi_c^{\text{points}})$, the coefficient of variation and the skewness of the distribution for networks with different mean coverages are shown in Figures 4.11 to 4.13. As might be anticipated, the variance passes through a maximum and the data are well described by a second-order polynomial of the form

$$\sigma^2(\Phi_c^{\text{points}}) = -k_1\,\bar{\Phi}_c^2 + k_2\,\bar{\Phi}_c \;,$$

where the coefficients k_1 and k_2 depend on the mean coverage and for higher coverages $k_1 \approx k_2 \approx 1/\bar{c}$. The coefficient of variation of fractional contact area at points decreases monotonically as does the skewness. As was observed qualitatively from Figure 4.10, we note from Figure 4.13 that the distribution is least symmetrical for high and low fractional contact areas, and close to symmetrical for a broader range of mean fractional contact areas when mean coverage is high, this being a consequence of the decreasing skewness of the Poisson distribution as the mean coverage increases.

We have shown that fractional contact area is a discrete random variable. The distribution of fractional contact area exhibits a positive skew for networks of low porosity and a negative skew for networks of high porosity, such that it is symmetrical when $\epsilon \approx \frac{1}{2}$; the maximum variance occurs at $\epsilon \approx \frac{1}{2}$ also. The influence of fibre surfaces is greatest for networks of low mean coverage.

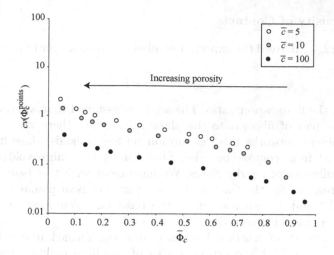

Figure 4.12. Collected results for the dependence of the coefficient of variation of fractional contact area at points, $cv(\Phi_c^{\mathrm{points}})$ on the mean fractional contact area, $\bar{\Phi}_c$ for networks with different mean coverage

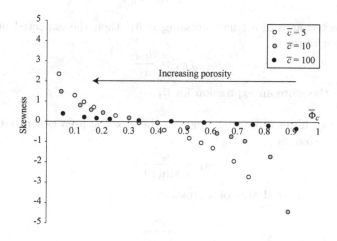

Figure 4.13. Collected results for the dependence of the skewness of the distribution of fractional contact area at points on the mean fractional contact area, $\bar{\Phi}_c$ for networks with different mean coverage

4.3.3 Intensity of Contacts

On page 112, we derived the expected number of crossings per fibre in a planar network:

$$\bar{n}_{\text{cross,fib}} = \frac{2}{\pi} A \bar{c} \tag{4.16}$$

where A is the fibre aspect ratio. The analysis assumed that wherever a projection of a pair of fibres onto the plane intersected, there was a crossing. For near-planar networks this assumption no longer holds, since fibres may be separated from contact by other fibres or by inter-fibre voids resulting from the influence of nearby fibres. We have dealt with this issue when deriving expressions for the fractional contact area of near-planar networks in Section 4.3.1 and we can use this as the basis for calculating the expected number of contacts between fibres in such networks.

For a network of fractional contact area Φ_c, formed from fibres with length λ and width ω, the expected area of any fibre making contact with other fibres is

$$\bar{a}_{\text{tot}} = 2\,\lambda\,\omega\,\Phi_c \tag{4.60}$$
$$= 2\,A\,\omega^2\,\Phi_c \ . \tag{4.61}$$

If the expected area of a single crossing is $\bar{a}_1$, then the expected number of crossings per fibres is

$$\bar{n}_{\text{cross,fib}} = \frac{\bar{a}_{\text{tot}}}{\bar{a}_1} \ . \tag{4.62}$$

We require therefore an expression for $\bar{a}_1$.

Where two fibres cross with angle ϕ, the crossing area generated is a parallelogram with area

$$a_1 = \frac{\omega^2}{\sin(\phi)} \ , \tag{4.63}$$

such that the expected area of a crossing is

$$\bar{a}_1 = \frac{\omega^2}{\sin(\phi)} \ . \tag{4.64}$$

Recall from Equation 4.11 that the probability density of crossing angles ϕ is

$$f(\phi) = \sin(\phi) \ .$$

We compute the probability density of the random variable $q = \sin(\phi)$ as

```
In[1]:=  f[ϕ_] := Sin[ϕ]
         ϕ = ArcSin[q];
         pdfq = f[ϕ] D[ϕ, q]
```

Out[3]= $\dfrac{q}{\sqrt{1-q^2}}$

So the expected value, $\bar{q} = \overline{\sin(\phi)}$ is,

In[4]:= **qbar = Integrate[q pdfq, {q, 0, 1}]**

Out[4]= $\dfrac{\pi}{4}$

We note that this differs from the case of fibres with a uniform distribution of orientations θ crossing a given line, where the expected value of $\sin(\theta)$ is $2/\pi$. Note also that the expected value of ϕ is $\bar{\phi} = 1$ and thus $\overline{\sin(\phi)} \neq \sin(\bar{\phi})$.

The expected area of a crossing is therefore

In[5]:= **a1 = ω^2 / qbar**

Out[5]= $\dfrac{4\,\omega^2}{\pi}$

and the expected number of crossings per fibre is

In[6]:= **atot = 2 A ω^2 Φc;**
ncross = atot / a1

Out[7]= $\dfrac{A\,\pi\,\Phi\mathrm{c}}{2}$

So the expected number of crossings per fibre increases linearly with aspect ratio and fractional contact area, as expected. To obtain the expected number of crossings per fibre in terms of network coverage and porosity, we use the expression for Φ_c derived in Section 4.3.1:

In[8]:= **Φ2d = 1 + (1 - ϵ) / Log[ϵ];**
Φinf = FullSimplify[Φ2d + (1 - ϵ)^2 (1 - Φ2d), 0 < ϵ < 1];
f = 1 + (EulerGamma - ExpIntegralEi[cbar] + Log[cbar]) /
(e^{cbar} - 1);
Φc = f Φinf;
ncross

Out[12]= $\frac{1}{2}$ A π $\left(1 + \dfrac{\text{EulerGamma} - \text{ExpIntegralEi}[\text{cbar}] + \text{Log}[\text{cbar}]}{-1 + e^{\text{cbar}}}\right)$

$\left(1 + \dfrac{\epsilon \ (2 + (-3 + \epsilon) \ \epsilon)}{\text{Log}[\epsilon]}\right)$

A three-dimensional plot of the expected number of crossings per fibre, in units of the aspect ratio, shows a non-linear decrease with increasing porosity and a non-linear increase with mean coverage $\bar{c}$:

In[13]:= **Plot3D[ncross /. A → 1, {ε, 0, 1},**
       {cbar, 0, 20}, AxesLabel → {"ε", "c̆", "ncross"}]

Out[13]=

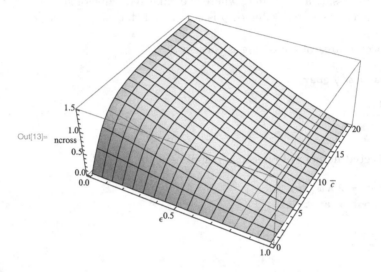

We note that the influence of porosity on the number of crossings per fibre is the same as that observed for the fractional contact area of networks with infinite thickness, Φ_∞. The mean coverage decreases the expected number of crossings per fibre of a network with infinite mean coverage by the fraction of the surface available for contact f.

The expected number of crossings per fibre is given by

$$n_{\text{cross}} = \frac{A}{2} \pi \Phi_c ,\qquad (4.65)$$

where A is the fibre aspect ratio and Φ_c is the fractional contact area of a network with mean coverage $\bar{c}$. The upper limit for the number of crossings per fibre is therefore about $\frac{3}{2}$ times the fibre aspect ratio and not 2 times, as would be given by the maximal packing of fibres with perpendicular orientation.

The treatment given here provides an alternative mechanism to estimate the percolation threshold of a fibre network, as discussed in Section 4.2. There we saw that from the simulations of Pike and Seager [125], Niskanen *et al.* [108] give the mean coverage at which a network of fibres with uniform length and width percolates as

$$\bar{c}_{\text{perc}} = \frac{5.7}{A} \, . \tag{4.9}$$

If the number of crossings per fibre is a Poisson variable with mean $\bar{n}_{\text{cross,fib}}$, then the fraction of fibres connected to no others is the Poisson probability that $n_{\text{cross,fib}} = 0$:

$$P(n_{\text{cross,fib}} = 0) = e^{-\bar{n}_{\text{cross,fib}}} \, . \tag{4.66}$$

If we assume that for percolation we require $P(n_{\text{cross,fib}} = 0) \leq 1\ \%$, then the expected number of crossings per fibre required for percolation is

In[14]:= **ncrossperc =**
ncrossbar /. Solve$\left[\text{e}^{\text{-ncrossbar}} == 0.01,\ \text{ncrossbar}\right]$[[1]]

Out[14]= 4.60517

The fractional contact area of the percolated network is therefore:

In[15]:= **Φ2dperc = ncrossperc al$\left/\left(2\,\text{A}\,\omega^2\right)\right.$**

Out[15]= $\dfrac{2.93174}{A}$

At the percolation threshold, we may consider our networks to be two-dimensional and we can therefore use Equation 4.24 for their fractional contact area. At very low mean coverages, we can approximate the fractional contact area by the first term of a Taylor series, as represented by the dashed line in the following plot:

In[16]:= **Φ2d = 1 + $\left(\text{e}^{\text{-cbar}} - 1\right)/\text{cbar}$;**
Φ2dApprox = Normal[Series[Φ2d, {cbar, 0, 1}]]
Plot[{Φ2d, Φ2dApprox}, {cbar, 0, 0.5},
PlotStyle → {{}, Dashed}, AxesLabel → {"c̄", "Φ$_{2d}$"}]

Out[17]= $\dfrac{\text{cbar}}{2}$

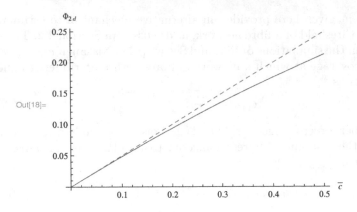

We now solve to determine the mean coverage at the percolation threshold:

In[19]:= **cbarperc =**
         Simplify[cbar /. Solve[Φ2dApprox == Φ2dperc, cbar][[1]]]

Out[19]= $\dfrac{5.86348}{A}$

The resultant expression agrees very closely with Equation 4.9.

4.3.4 Absolute Contact States

At any given point in the plane of a network, a fibre covering that point may be in contact with 1, 2 or no other fibres above or below it, no other configuration being possible. These *absolute contact states* are important because we expect regions where there is no contact to exhibit different stress-strain behaviours from those with one or two contacts. Similarly, we might expect the fracture path through a sheet to be influenced to some extent by the configuration of fibre contacts and the compression behaviour of a given fibre segment within a sheet to be determined by the local number and configuration of contacts and those of nearby regions.

Kallmes *et al.* [76] considered the distribution of absolute contact states in two-dimensional networks such that every fibre crossing was assumed to generate a contact and the probability of coverage greater than three was negligible. The fraction of the total fibrous length which does not make contacts with other fibres is that with coverage 1, divided by the mean coverage:

In[1]:= **P[c_] := PDF[PoissonDistribution[cbar], c]**
        C0 = P[1] / cbar

Out[2]= e^{-cbar}

At points with coverage greater than 1, the two outermost fibres are in contact with other fibres on one side only and this fraction is

In[3]:= **C1 = Factor[2 Sum[P[c] / cbar, {c, 2, ∞}]]**

Out[3]=
$$\frac{2\,e^{-cbar}\left(-1 - cbar + e^{cbar}\right)}{cbar}$$

Similarly, at points with coverage greater than 2, all except the two outermost fibres make contact on both sides:

In[4]:= **C2 = FullSimplify[Sum[(c - 2) P[c] / cbar, {c, 3, ∞}]]**

Out[4]=
$$\frac{e^{-cbar}\left(2 + cbar + (-2 + cbar)\,e^{cbar}\right)}{cbar}$$

such that $C_0 + C_1 + C_2 = 1$:

In[5]:= **TrueQ[Simplify[C0 + C1 + C2] == 1]**

Out[5]= True

Kallmes *et al.* presented these expressions in a representation referred to as the 'Bonding State Diagram' which we term the 'contact state diagram' instead. The diagram is produced by plotting C_0, C_1 and C_2 against the fractional contact area for two-dimensional networks as given on page 115:

$$\Phi_{2D} = 1 + \frac{e^{-\bar{c}} - 1}{\bar{c}}, \tag{4.67}$$

such that the only variable used to generate these curves is the mean coverage of the network. The expressions provided by Kallmes *et al.* [76] for C_0, C_1 and C_2, and derived above apply only to networks with mean coverage less than about 1 and, from Equation 4.67, this corresponds to $\Phi_{2D} \approx 0.4$. Accordingly, we plot the contact state diagram outside this range using broken lines.

```
In[6]:= Φ2D = 1 + (e^-cbar - 1)/cbar;
       cb = cbar /. Solve[Φ2D == fca, cbar][[1]];
       Show[Plot[{1 - C1 /. cbar → cb, C0 /. cbar → cb},
         {fca, 0, 1}, PlotStyle → None,
         Filling → {1 → {2}}, FillingStyle → LightGray],
       ParametricPlot[{{Φ2D, C0}, {Φ2D, 1 - C1}},
         {cbar, 10^-10, 1}, PlotStyle → Black],
       ParametricPlot[{{Φ2D, C0}, {Φ2D, 1 - C1}},
         {cbar, 1, 1000}, PlotStyle → {{Black, Dashed}}],
       AspectRatio → 1, Frame → True,
       FrameTicks → {Automatic, Automatic, None, {{1, 0},
         {.8, .2}, {.6, .4}, {.4, .6}, {.2, .8}, {0, 1}}},
       FrameLabel → {"Φ2D", "C0", None, "C1"}, PlotRange → All]
```

Out[8]=

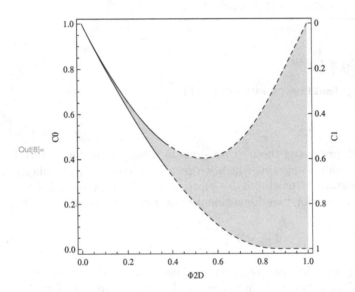

In this graphical representation, the shaded region corresponds to the fraction C_2, the region above this corresponds to the fraction C_1 and the region below to C_0. As we have seen, in real networks porosity and hence fractional contact area can be varied independently of coverage. Despite this, and the fact that the theory is limited to low coverages only, Kallmes *et al.* observed reasonable agreement between their theory and data generated in a Monte Carlo simulation where rectangular fibres of constant length and different aspect ratio, A were placed in a rectangular computational grid; the coordinates of fibre centres and the orientation of the major axes of fibres being generated as uniformly distributed random numbers. These data are included in the contact state diagram shown in Figure 4.14 where the solid lines represent the absolute contact states for two-dimensional networks. The broken lines repre-

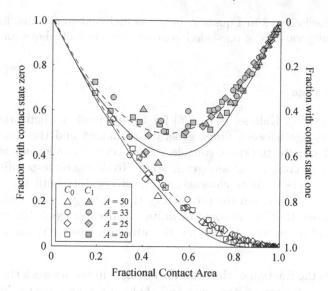

Figure 4.14. The contact state diagram after Kallmes *et al.* [76] including comparison with simulation data for fibres of different aspect ratio, A; broken lines approximate the multi-planar model given in [76]

sent second-order polynomial regressions on the data and approximate theory provided by Kallmes *et al.* for the contact states of fibres in multi-planar networks formed by the superposition of two-dimensional networks with a given probability of contact occurring between pairs of layers obscured by a third. The expressions resulting from this treatment are omitted here, since we shall obtain further results from a simplified treatment in the sequel; they are provided however in the appendix to [76] along with the data used to generate Figure 4.14.

Infinite Mean Coverage

Consider first a random fibre network of infinite mean coverage. The fraction of fibre surfaces contacting other fibres is the fractional contact area, Φ_∞ such that a fraction $(1 - \Phi_\infty)$ of fibre surfaces is not in contact with other fibres. At any given point in the network, a given fibre may be in contact with 1, 2 or no other fibres above or below it with the probability of contact on each side of the fibre being Φ_∞. It follows that absolute contact states for networks of infinite mean coverage are given by

$$C_0^\infty = (1 - \Phi_\infty)^2 \;, \tag{4.68}$$

$$C_1^\infty = 2\Phi_\infty (1 - \Phi_\infty) \;, \tag{4.69}$$

$$C_2^\infty = \Phi_\infty^2 \;, \tag{4.70}$$

where the coefficient 2 in Equation (4.69) is included to account for the two possible configurations of one-sided contact, *i.e.* contact above and below a given fibre.

Finite coverage

Now, the model of Kallmes *et al.* [76] for two-dimensional networks assumes that all crossings between fibres generate a contact and the treatment for networks of infinite thickness considers fibre contacts independently of the number of fibre crossings at any given point. To derive corresponding expressions for networks of finite coverage, we must consider both the number of crossings at any point and the probability that these generate a contact.

It turns out that in networks of finite mean coverage $\bar{c}$, the expressions of Kallmes *et al.* provide bounds on the absolute contact states of the network [144]:

- C_0 is the fraction of the total fibre length in the network that may only have contact state zero and that cannot have contact state 1 or 2;
- C_1 is the fraction of the total fibre length in the network that may have contact state zero or 1 but that cannot have contact state 2;
- C_2 is the fraction of the total fibre length in the network that may have any of the three permissible contact states.

We consider contacts at points only, so the cross-sectional geometry of fibres, their flexibility, and hence their propensity to conform to each other, are not an issue. We assume that the probability that a given point on a fibre's surface is in contact with another fibre is Φ_∞ and that it is independent of the coverage at that point. The fraction of C_1 that makes contact with other fibres is

$$\Phi_\infty C_1 ,$$

and the fraction which does not make contact is

$$(1 - \Phi_\infty) C_1 .$$

Similarly, the fraction of C_2 that makes contact with one other fibre is

$$2 (1 - \Phi_\infty) \Phi_\infty C_2 .$$

The fraction that makes contact with two other fibres is

$$\Phi_\infty^2 C_2 ,$$

and the fraction which does not make contact is

$$(1 - \Phi_\infty)^2 \, C_2 \; .$$

Accordingly, the fractions of the network with contact states zero, one and two are given by

$$C_0^c = C_0 + (1 - \Phi_\infty) \, C_1 + (1 - \Phi_\infty)^2 \, C_2 \; , \tag{4.71}$$

$$C_1^c = \Phi_\infty \, C_1 + 2 \, \Phi_\infty \, (1 - \Phi_\infty) \, C_2 \; , \tag{4.72}$$

$$C_2^c = \Phi_\infty^2 \, C_2 \; , \tag{4.73}$$

respectively.

We input these expressions to *Mathematica* and use **Factor** to obtain a simplified form when appropriate.

```
In[9]:=  Cc0 = Factor[C0 + (1 - Φinf) C1 + (1 - Φinf)² C2]
         Cc1 = Factor[Φinf C1 + 2 Φinf (1 - Φinf) C2]
         Cc2 = Φinf² C2
```

$$\text{Out[9]}= \frac{1}{\text{cbar}} e^{-\text{cbar}} \big(\text{cbar } e^{\text{cbar}} - 2 \, \Phi\text{inf} + 2 \, e^{\text{cbar}} \, \Phi\text{inf} - 2 \, \text{cbar } e^{\text{cbar}} \, \Phi\text{inf} +$$
$$2 \, \Phi\text{inf}^2 + \text{cbar } \Phi\text{inf}^2 - 2 \, e^{\text{cbar}} \, \Phi\text{inf}^2 + \text{cbar } e^{\text{cbar}} \, \Phi\text{inf}^2 \big)$$

$$\text{Out[10]}= -\frac{1}{\text{cbar}} 2 \, e^{-\text{cbar}} \, \Phi\text{inf} \big(-1 + e^{\text{cbar}} - \text{cbar } e^{\text{cbar}} +$$
$$2 \, \Phi\text{inf} + \text{cbar } \Phi\text{inf} - 2 \, e^{\text{cbar}} \, \Phi\text{inf} + \text{cbar } e^{\text{cbar}} \, \Phi\text{inf} \big)$$

$$\text{Out[11]}= \frac{e^{-\text{cbar}} \big(2 + \text{cbar} + (-2 + \text{cbar}) \, e^{\text{cbar}} \big) \, \Phi\text{inf}^2}{\text{cbar}}$$

```
In[12]:=  Simplify[Cc0 + Cc1 + Cc2]
```

```
Out[12]=  1
```

Of course, when dealing with networks of finite coverage $\bar{c}$, we require the fractional contact states to be expressed in terms of the fractional contact area $\Phi_c = \Phi_\infty f$ where f is the fraction of the network available for contact with other fibres as discussed on page 134:

```
In[13]:=  f = FullSimplify[
            1 / (1 - P[0]) Sum[(c - 1) / c P[c], {c, 2, ∞}], cbar > 0]
          Φinf = Φc / f;
```

$$\text{Out[13]}= 1 + \frac{\text{EulerGamma} - \text{ExpIntegralEi}[\text{cbar}] + \text{Log}[\text{cbar}]}{-1 + e^{\text{cbar}}}$$

We note that in the limit as $\Phi_c \to f$, we recover the original expressions of Kallmes *et al.* [76]:

In[15]:= **Limit[Cc0, Φc → f]**
Limit[Cc1, Φc → f]
FullSimplify[Limit[Cc2, Φc → f]]

Out[15]= e^{-cbar}

Out[16]= $\dfrac{2\, e^{-cbar}\, \left(-1 - cbar + e^{cbar}\right)}{cbar}$

Out[17]= $\dfrac{e^{-cbar}\, \left(2 + cbar + (-2 + cbar)\, e^{cbar}\right)}{cbar}$

and in the limit as $\bar{c} \to \infty$, we recover Equations 4.68 to 4.70:

In[18]:= **Limit[Cc0, cbar → ∞]**
Limit[Cc1, cbar → ∞]
Limit[Cc2, cbar → ∞]

Out[18]= $(-1 + \Phi c)^2$

Out[19]= $-2\,(-1 + \Phi c)\,\Phi c$

Out[20]= Φc^2

We may now plot a new contact state diagram showing the influence of coverage and fractional contact area on the absolute contact states. We demonstrate this for networks of mean coverage $\bar{c} = 5$ and $\bar{c} = 10$ and plot the case for networks of infinite coverage using dashed lines.

```
In[21]:= GraphicsArray[
    {{Show[Plot[{1 - Cc1 /. cbar → 5, Cc0 /. cbar → 5},
        {Φc, 0, f /. cbar → 5}, Filling → {1 → {2}},
        FillingStyle → LightGray],
      Plot[{(1 - Φc)², 1 - (2 Φc (1 - Φc))}, {Φc, 0, 1},
        PlotStyle → Dashed], AspectRatio → 1, Frame → True,
      FrameTicks → {Automatic, Automatic, None, {{1, 0},
          {.8, .2}, {.6, .4}, {.4, .6}, {.2, .8}, {0, 1}}},
      FrameLabel → {"Φc", "C₀ᶜ", None, None},
      PlotRange → {{0, 1}, {0, 1}}, PlotLabel → "c̄=5"],
    Show[Plot[{1 - Cc1 /. cbar → 10, Cc0 /. cbar → 10},
        {Φc, 0, f /. cbar → 10}, Filling → {1 → {2}},
        FillingStyle → LightGray],
      Plot[{(1 - Φc)², 1 - (2 Φc (1 - Φc))}, {Φc, 0, 1},
        PlotStyle → Dashed], AspectRatio → 1, Frame → True,
      FrameTicks → {Automatic, Automatic, None, {{1, 0},
          {.8, .2}, {.6, .4}, {.4, .6}, {.2, .8}, {0, 1}}},
      FrameLabel → {"Φc", None, None, "C₁ᶜ"},
      PlotRange → {{0, 1}, {0, 1}}, PlotLabel → "c̄=10"]}}]
```

Out[21]=

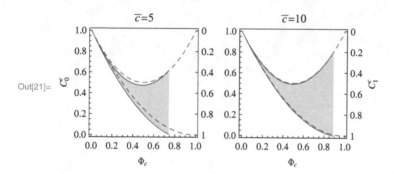

We observe that at mean coverages less than about 10, $C_0^c < C_0^\infty$ and $C_1^c > C_1^\infty$, the net effect, not shown here, being that $C_2^c > C_2^\infty$ also. For mean coverages of 10 or higher, the influence of the surfaces on the absolute contact states is minimal, as shown in the plot on the right. Note however that range of Φ_c remains truncated via the influence of our parameter f.

5

Poisson Fibre Processes II: Void Phase

5.1 Introduction

In the previous chapter, we derived expressions for the fractional contact area, and hence the expected number of crossings with network porosity as the free parameter. We have seen also that for networks of infinite thickness, fibre contact and the absolute contact states depend on porosity only. For networks of finite thickness, variables characterising fibre morphology must be considered, along with the mean areal density in order that we may account for surface effects through the dimensionless variable coverage. Evidently, porosity is an important structural variable and impacts strongly on other structural features of the network. Recall also that we may capture the influence of fibre flexibility on network properties through its influence on porosity, simplifying many analyses.

In addition to its strong influence on other properties, network porosity is an important structural parameter in its own right; in particular the organisation of the void space will control the suitability of a given network for certain applications. Inevitably, if a network is to be used as a filter then we require its pore size distribution to be compatible with the distribution of diameters of the particles we seek to capture. It is well established also that the permeability of networks is influenced not only by their porosity, but by their pore size distribution also and there is recent evidence that the ingress of cells into nanofibrous scaffolds for tissue engineering is strongly dependent on pore size distribution [121].

The concept of a pore size distribution is intuitively simple for thin networks such as those discussed in Section 3.3.3. For such networks, polygonal voids exist as discrete entities with measurable area and perimeter, whereas for thicker networks the void space is highly interconnected and tortuous. Classical methods of measuring pore size and its distribution, such as mercury porosimetry and capillary flow porometry [70, 93] associate a fractional volume or flow with the dimensions of constrictions in a path through the material to generate a pore diameter distribution. Thus, when modelling pore size distributions there are two primary characteristics of interest; firstly, the global distribution of void dimensions, and secondly, the distribution of the dimensions of the narrowest point in a path through the network.

We concern ourselves here with multi-planar structures formed by superimposing planar networks. For these materials, voids will typically have different dimensions when measured in the plane to those measured perpendicular to it. Our treatment therefore considers first the in-plane dimensions of voids and proceeds to consider their out-of-plane dimensions. Having established models for these two properties, we combine them to characterise the distribution of the narrowest dimensions encountered in a path from one side of the network to the other.

For an in-plane measure of pore size, such as a hydraulic radius or equivalent pore diameter, we may assume the theory for two-dimensional networks to hold, so, from the treatment of Corte and Lloyd [20] discussed in Section 3.3.3 we expect the distribution of equivalent pore diameters in a random network to have a gamma distribution with coefficient of variation $cv(d) = \sqrt{16 - \pi^2}/\pi$.

5.2 In-plane Pore Dimensions

Corte and Lloyd [20] give the expected pore diameter of a two-dimensional fibre network as

$$\bar{d} = \frac{\sqrt{\pi}}{2}\,\bar{g}\ ,$$

where $\bar{g}$ is the expected inter-crossing distance. For a fibre of length λ we may compute $\bar{g}$ and hence $\bar{d}$ from the expected number of crossings per fibre $\bar{n}_{\mathrm{cross,fib}}$,

$$\bar{g} = \frac{\lambda}{\bar{n}_{\mathrm{cross,fib}} - 1}$$

$$\approx \frac{\lambda}{\bar{n}_{\mathrm{cross,fib}}} \qquad \text{for } \bar{n}_{\mathrm{cross,fib}} \gg 1. \tag{5.1}$$

More precisely, Equation 5.1 gives the expected distance between the points where the principal axes of fibres cross a given fibre. For fibres of width ω, the expected inter-crossing distance will be smaller by an amount $\omega/\sin(\phi) = 4\omega/\pi$. Recall from Equation 4.16 the expected number of crossings per fibre is,

$$\bar{n}_{\mathrm{cross,fib}} = \frac{2\,\lambda}{\pi\,\omega}\,\bar{c}\ ,$$

so we have,

$$\bar{g} \approx \left(\frac{\pi}{2\,\bar{c}} - \frac{4}{\pi}\right)\omega$$

$$\approx \left(\frac{\pi}{2\log(1/\epsilon)} - \frac{4}{\pi}\right)\omega\ , \tag{5.2}$$

where $\epsilon = e^{-\bar{c}}$ is the fractional open area of the two-dimensional network.

Equation 5.2 tells us that the expected inter-fibre crossing distance and hence the expected pore radius depends on fractional open area and fibre width only. At mean coverages $\bar{c} > \pi^2/8 \approx 0.8$ however, it yields negative values of $\bar{g}$ since the model is unable to distinguish between crossings where fibre axes are sufficiently close that fibres generating adjacent crossings overlap. An alternative approach is to use the assumption that the expected number of crossings per unit area is the same as the expected number of polygonal voids per unit area; from Equation 4.15 we have therefore,

$$\bar{n}_{\text{void}} = \bar{n}_{\text{cross}} = \frac{\bar{c}^2}{\pi\,\omega^2}$$

$$= \frac{\log^2(1/\epsilon)}{\pi\,\omega^2} . \tag{5.3}$$

Recall the result of Miles [102] that for a random line process with intensity τ, there is no influence of fibre width on the mean polygon area or the distribution of polygon areas. We may therefore use Equation 5.3 to obtain the expected area of a polygonal void without being concerned that voids may not be generated where fibre crossings are very close together. The expected area of a polygonal void is therefore,

$$\bar{a}_{\text{void}} = \frac{1}{\bar{n}_{\text{void}}}$$

$$= \frac{\pi\,\omega^2}{\log^2(1/\epsilon)} \tag{5.4}$$

It follows that the expected equivalent pore radius is,

$$\bar{d} = 2\sqrt{\frac{\bar{a}_{\text{void}}}{\pi}}$$

$$= \frac{2\,\omega}{\log(1/\epsilon)} . \tag{5.5}$$

Again, we see that the expected pore diameter of a two-dimensional network depends only upon the fibre width and its fractional open area. Equation 5.5 is valid over the full range of ϵ however, because the influence of fibre width was accounted for in the derivation of the expected number of crossings per unit area. A plot of the mean pore radius in units of the fibre width, as given by Equation 5.5, shows the mean pore diameter to be extremely sensitive to porosity at high and low porosities, and to exhibit an approximately log-linear dependence for porosities between about 0.2 and 0.7:

```
In[1]:= dbar = 2 ω / Log[1 / ε];
        LogPlot[dbar / ω, {ε, 0, 1}, AxesLabel → {"ε", "d̄/ω"}]
```

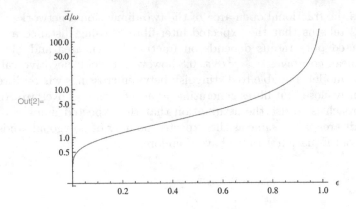

Equation 5.5 gives the mean pore diameter of two-dimensional networks with constant porosity as being proportional to fibre width. Of course, if the total fibre length per unit area remains constant, and fibre width increases, then the porosity will decrease, and so will the mean pore diameter.

For multi-planar structures with porosity ϵ, the in-plane dimensions of voids are also given by Equation 5.5. Measurements of the mean pore diameter of such materials show this to decrease with increasing areal density [9, 20, 146] and similarly filtration efficiency improves [93]. With increasing areal density, the length of paths from one side of the network to the other increases also and hence so does the probability of encountering a small pore, this being what is measured when determining pore size or filtration efficiency. To account for this in a model, we must consider the multi-planar structures formed by the superimposition of n two-dimensional layers and follow the treatment in [142].

Consider a layered structure of circular voids with gamma distributed diameters. We input the probability density and cumulative distribution functions into *Mathematica* as,

In[3]:= **pdfd = PDF[GammaDistribution[α, β / α], d]**
cdfd =
FunctionExpand[CDF[GammaDistribution[α, β / α], d]]

Out[3]= $\dfrac{d^{-1+\alpha}\, e^{-\frac{d\alpha}{\beta}} \left(\frac{\beta}{\alpha}\right)^{-\alpha}}{\mathrm{Gamma}[\alpha]}$

Out[4]= $1 - \dfrac{\mathrm{Gamma}\left[\alpha,\, \frac{d\alpha}{\beta}\right]}{\mathrm{Gamma}[\alpha]}$

A second layer with an independent and identical distribution of pore diameters is placed over the first layer such that the centres of pairs of voids in the two layers are aligned. For such a structure, we assign to each pair of pores the diameter of the smaller pore. The probability that a pore of diameter d is larger than another pore drawn at random from the same distribution is given by the cumulative distribution function for d, so the probability density of diameters of the smallest of each pair of pores is given by,

In[5]:= **pdf2layers = 2 (1 - cdfd) pdfd**

$$\text{Out[5]=} \quad \frac{2\, d^{-1+\alpha}\, e^{-\frac{d\alpha}{\beta}}\, \left(\frac{\beta}{\alpha}\right)^{-\alpha}\, \text{Gamma}\left[\alpha,\, \frac{d\alpha}{\beta}\right]}{\text{Gamma}[\alpha]^2}$$

and the cumulative distribution function for our two layer structure is,

In[6]:= **cdf2layers =**
Simplify[PowerExpand[Integrate[pdf2layers,
{d, 0, d}, Assumptions → Re[α] > 0]]]

$$\text{Out[6]=} \quad 1 - \frac{\text{Gamma}\left[\alpha,\, \frac{d\alpha}{\beta}\right]^2}{\text{Gamma}[\alpha]^2}$$

Consider now the addition of a third layer. The probability that a pore of diameter d drawn from our two layer structure is larger than another pore drawn at random from the third layer is given by the cumulative distribution function for d of the two layer structure, so the probability density of diameters of the smallest of each pair of pores is given by,

In[7]:= **pdf3layers = 3 (1 - cdf2layers) pdfd**

$$\text{Out[7]=} \quad \frac{3\, d^{-1+\alpha}\, e^{-\frac{d\alpha}{\beta}}\, \left(\frac{\beta}{\alpha}\right)^{-\alpha}\, \text{Gamma}\left[\alpha,\, \frac{d\alpha}{\beta}\right]^2}{\text{Gamma}[\alpha]^3}$$

and the cumulative distribution function for our three layer structure is,

In[8]:= **cdf3layers =**
Simplify[PowerExpand[Integrate[pdf3layers,
{d, 0, d}, Assumptions → Re[α] > 0]]]

$$\text{Out[8]=} \quad 1 - \frac{\text{Gamma}\left[\alpha,\, \frac{d\alpha}{\beta}\right]^3}{\text{Gamma}[\alpha]^3}$$

Similarly, for a four layer structure:

In[9]:= **pdf4layers = 4 (1 - cdf3layers) pdfd**
cdf4layers =
  Simplify[PowerExpand[Integrate[pdf4layers,
    {d, 0, d}, Assumptions → Re[α] > 0]]]

Out[9]=
$$
\frac{4\ d^{-1+\alpha}\ e^{-\frac{d\alpha}{\beta}}\ \left(\frac{\beta}{\alpha}\right)^{-\alpha}\ \mathrm{Gamma}\!\left[\alpha,\ \frac{d\alpha}{\beta}\right]^{3}}{\mathrm{Gamma}[\alpha]^{4}}
$$

Out[10]=
$$
1 - \frac{\mathrm{Gamma}\!\left[\alpha,\ \frac{d\alpha}{\beta}\right]^{4}}{\mathrm{Gamma}[\alpha]^{4}}
$$

Evidently, a pattern is emerging. For a multi-planar structure consisting of n layers the distribution of the smallest of n pores has probability density:

In[11]:= **pdf[n_, d] := n $\left((1 - cdfd)^{n-1}\right)$ pdfd**
pdf[n, d]

Out[12]=
$$
\frac{d^{-1+\alpha}\ e^{-\frac{d\alpha}{\beta}}\ n\ \left(\frac{\beta}{\alpha}\right)^{-\alpha}\ \left(\dfrac{\mathrm{Gamma}\!\left[\alpha,\ \frac{d\alpha}{\beta}\right]}{\mathrm{Gamma}[\alpha]}\right)^{-1+n}}{\mathrm{Gamma}[\alpha]}
$$

and cumulative distribution function:

In[13]:= **cdf[n_, d] := 1 - (1 - cdfd)^{n}**
cdf[n, d]

Out[14]=
$$
1 - \left(\frac{\mathrm{Gamma}\!\left[\alpha,\ \frac{d\alpha}{\beta}\right]}{\mathrm{Gamma}[\alpha]}\right)^{n}
$$

We have parametrised our distribution in terms of the mean pore radius of a given layer, $\bar{d} = \beta$ and its coefficient of variation, which is known for random networks from the result of Corte and Lloyd [20] derived on page 92, $cv(d) = \sqrt{16 - \pi^2}/\pi = 1/\sqrt{\alpha}$. Making these assignments in *Mathematica* we can plot the surface showing the influence of the number of layers n on the probability density of pore diameters in units of the fibre width. Our example is generated for a network with porosity $\epsilon = 0.5$.

```
In[15]:=  β = dbar / ω;
          α = 1 / cvd²;

          cvd = √(16 - π²) / π;

          ε = .5;
          Plot3D[Evaluate[pdf[n, d]], {d, 0, 6}, {n, 1, 10},
             PlotRange → All, AxesLabel → {"d/ω", "n"}]
          ε = .
```

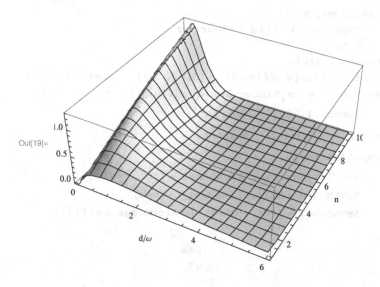

We observe that the distribution is shifted towards smaller pore radii as the number of layers increases. As we saw when modelling fractional contact area in Section 4.3.1, the number of layers required to model a network of mean coverage $\bar{c}$ is determined by the network porosity,

$$n = \frac{\bar{c}}{\bar{c}_{2D}} = \frac{\bar{c}}{\log(1/\epsilon)} \, . \tag{4.51}$$

To determine the mean, variance and coefficient of variation of the multi-planar pore diameter d_{mp}, requires numerical integration:

```
In[21]:=  Clear[β]
          n = cbar / Log[1 / ε]; β = dbar / ω;
          ε = 0.5; cbar = 5;
          dbarmp = NIntegrate[d pdf[n, d], {d, 0, ∞}]
          dvarmp = NIntegrate[(d - dbarmp)² pdf[n, d], {d, 0, ∞}]

          cvdmp = √dvarmp / dbarmp
```

Out[24]= 0.477555

Out[25]= 0.112007

Out[26]= 0.700809

To investigate the influence of mean coverage on the pore diameter distribution, we perform multiple integrations within a **Table** environment:

```
In[27]:= Clear[cbar, n, β]
        n = cbar / Log[1 / ε]; β = dbar / ω;
        ε = 0.5;
        dbartab = Table[
            NIntegrate[d pdf[n, d] , {d, 0, ∞}], {cbar, 1, 50}];
        dvartab = Table[NIntegrate[d² pdf[n, d] , {d, 0, ∞}],
            {cbar, 1, 50}] - dbartab²;
        cvdtab = √dvartab / dbartab;
        GraphicsArray[
          {{ListPlot[dbartab, AxesLabel → {"c̄", "d̄/ω"},
            PlotRange → All], ListPlot[cvdtab,
            AxesLabel → {"c̄", "cv(d)"}, PlotRange → All]}}]
```

Out[33]=

When considering the influence of fibre variables, we must ensure that we take account of the influence of these on the mean coverage of the network; similarly, when considering the influence of porosity, we must take account of its influence on the number of layers required to generate a multi-planar structure. The following example computes the influence of fibre width on the mean pore diameter of a random fibre network with mean areal density $\bar{\beta} = 20$ g m^{-2} formed with porosity $\epsilon = 0.5$ from nylon fibres with circular cross section and density 1,100 kg m^{-3}. Care must be taken when handling the units of these variables and these are included as comments in the code. Here we examine the influence of changing fibre width 1 μm $\leq \omega \leq$ 20 μm:

```
In[34]:= Clear[cbar, n, β, ε]
         n = cbar / Log[1 / ε];  β = dbar;
         βbar = 20 (* g m⁻² *);
         ρ = 1.1 10⁶ (* g m⁻³ *);
         δ = ρ π ω² / 4 (* g m⁻¹ *);
         cbar = 10⁶ βbar ω / δ;
         ε = 0.5;
         ωdmp = Table[{ω, NIntegrate[d pdf[n, d], {d, 0, ∞}]},
             {ω, 1, 20}];
         ListPlot[ωdmp, AxesLabel → {"ω (μm)", "d̄ (μm)"}]
```

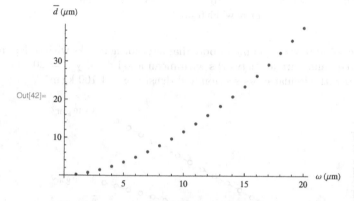

We observe that increasing fibre width for networks of constant porosity and areal density increases the mean in-plane pore diameter. This result arises because as we increase fibre width we must reduce the total fibre length per unit volume in order to occupy the same volume of the network with fibres and maintain its areal density.

> We state then, that the mean pore diameter is controlled by the total fibre length per unit volume which stochastically partitions the space occupied by the network.

Collected results showing the influence of network porosity, mean areal density, fibre width and fibre coarseness are given in Figures 5.1 to 5.4. Fibre width and fibre density both influence the total fibre length per unit volume at a given areal density, so pore diameter increases as each of these fibre properties increase. The influence of areal density is greatest at low areal densities, so above a certain weight a manufacturer of a fibrous filter, for example, will find it more cost effective to reduce pore size by influencing

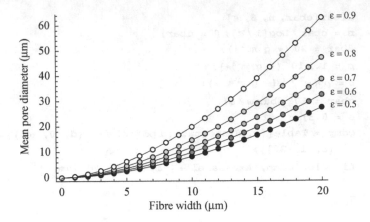

Figure 5.1. Effect of fibre width on mean pore diameter for networks with different porosity. Calculated values are for networks with mean areal density $\bar{\beta} = 20\ \mathrm{g\,m^{-2}}$ formed from fibres with circular cross section and density $\rho = 1,100\ \mathrm{kg\,m^{-3}}$

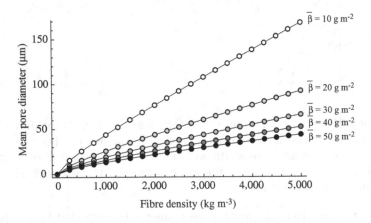

Figure 5.2. Effect of fibre density on mean pore diameter for networks with different areal density. Calculated values are for networks with porosity $\epsilon = 0.5$ formed from fibres with circular cross section with width $\omega = 20\ \mu\mathrm{m}$

porosity or fibre width than by making a heavier filter. The sensitivity of the mean pore diameter at high network porosities that we observed for two-dimensional networks persists in the multi-planar case, though in the range $0.2 < \epsilon < 0.7$, we observe an approximately linear dependence of mean pore diameter on porosity. Recent experimental data for electrospun fibre networks with high porosity show good agreement with mean pore diameters calculated using the model presented here [121].

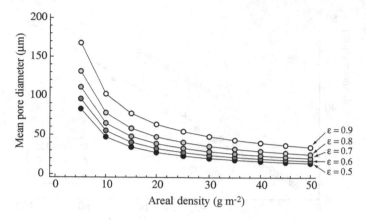

Figure 5.3. Effect of network grammage on mean pore diameter for networks with different porosity. Calculated values are for networks formed from fibres with circular cross section with width $\omega = 20$ µm and density $\rho = 1,100$ kg m^{-3}

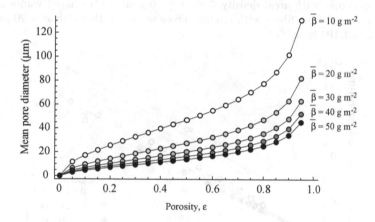

Figure 5.4. Effect of network porosity on mean pore diameter for networks with different areal density. Calculated values are for networks formed from fibres with circular cross section with width $\omega = 20$ µm and density $\rho = 1,100$ kg m^{-3}

Experimental measures of the pore diameter distribution in paper show the standard deviation to be approximately proportional to the mean [9, 20, 35, 141, 146]. Figures 5.5 and 5.6 show that the model captures this behaviour also. An important observation arising from this linear approximation is that the coefficient of variation of pore diameters can be considered to be approximately constant for changes in areal density and porosity [141]. We discussed earlier the result of Hwang and Hu [69] that for independent positive random

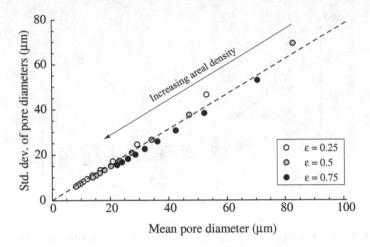

Figure 5.5. Standard deviation of pore diameter plotted against mean pore diameter for networks with areal density $5 \leq \bar{\beta} \leq 50$ g m^{-2}. Calculated values are for networks formed from fibres with circular cross section with width $\omega = 20$ μm and density $\rho = 1{,}100$ kg m^{-3}

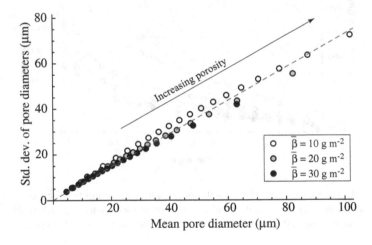

Figure 5.6. Standard deviation of pore diameter plotted against mean pore diameter for networks with porosity $0.05 \leq \epsilon \leq 0.95$. Calculated values are for networks formed from fibres with circular cross section with width $\omega = 20$ μm and density $\rho = 1{,}100$ kg m^{-3}

variables with a common continuous probability density, the property of the sample mean and coefficient of variation being independent is equivalent to the random variables being drawn from a gamma distribution. Assigning random porosity and coverage to our network, we observe that the distribution arising from a multi-planar structure with gamma distributed pore diameters in the constituent layers (solid line), itself closely resembles a gamma distribution with the same mean and variance (dashed line):

```
In[43]:= Clear[cbar, n, β, ε, βbar, ρ, ω]
         SeedRandom[1]
         n = cbar / Log[1 / ε]; β = dbar / ω;
         ε = RandomReal[];
         cbar = RandomReal[{0, 50}];
         dbarmp = NIntegrate[d pdf[n, d] , {d, 0, ∞}];
         dvarmp =
           NIntegrate[(d - dbarmp)² pdf[n, d] , {d, 0, ∞}];
         cvdmp = √dvarmp / dbarmp;
         Plot[Evaluate[{cdf[n, d], CDF[
             GammaDistribution[1 / cvdmp², dbarmp cvdmp²], d]}],
           {d, 0, 5 dbarmp}, PlotStyle → {{}, Dashed},
           AxesLabel → {"d", "cdf(d)"}]
```

5.3 Out-of-plane Pore Dimensions

So far we have considered only the dimensions of pores as measured in the plane of the network. When considering the fractional contact area of multi-planar networks, we noted that vertically adjacent fibres may not generate a contact due to the influence of nearby fibres. The vertical separation between pairs of such fibres is termed the pore height and provides another aspect of the dimensions of voids. The pore height distribution influences the compressibility of a fibre network and we expect it to influence its barrier properties also. We begin by deriving the mean pore height developing the treatment provided by Niskanen and Rajatora [109].

Consider first a network formed from fibres with mean thickness, *i.e.* dimension perpendicular to the plane of the network, $\bar{t}$. For a network with mean coverage $\bar{c}$, the expected number of interfaces between fibres at any point in the plane of the network is $(\bar{c}-1)$. Denoting the mean pore height $\bar{h}$, the expected total height of pores at any point in the network is

$$\bar{h}_{\text{tot}} = (\bar{c} - 1)\,\bar{h} \; , \tag{5.6}$$

such that the expected thickness of the network $\bar{z}$ is

$$\bar{z} = (\bar{c} - 1)\,\bar{h} + \bar{c}\bar{t} \; , \tag{5.7}$$

and network porosity is given by

$$\epsilon = \frac{\bar{h}_{\text{tot}}}{\bar{z}}$$
$$= \frac{(\bar{c} - 1)\,\bar{h}}{(\bar{c} - 1)\,\bar{h} + \bar{c}\bar{t}} \; . \tag{5.8}$$

Solving for $\bar{h}$ we obtain

```
In[1]:=  zbar = hbar (cbar - 1) + cbar tbar;
         htotbar = hbar (cbar - 1);
         hbar = hbar /. Solve[ε == (htotbar / zbar), hbar][[1]]
```

```
Out[3]=  -  cbar tbar ε
            ─────────────────────
            (-1 + cbar) (-1 + ε)
```

Niskanen and Rajatora [109] considered only the case where $\bar{c} \gg 1$ so we have,

$$\bar{h} \approx \frac{\epsilon}{1 - \epsilon}\,\bar{t} \; . \tag{5.9}$$

Now, by assuming the expected number of interfaces between fibres to be $(\bar{c}-1)$, the expression that we have obtained for mean pore height includes pores of height zero, *i.e.* fibre contacts. To exclude fibre contacts from our calculation of mean pore height we introduce the fractional contact area as a variable. The probability that a pair of vertically adjacent fibres make contact is the same as the fractional contact area of an infinitely thick network, Φ_∞, so the expected number of interfaces between fibres at any point in the plane of the network is $(\bar{c}-1)\,(1-\Phi_\infty)$. Incorporating this in our derivation we now have,

```
In[4]:=  zbar = hbarΦ (cbar - 1) (1 - Φinf) + cbar tbar;
         htotbar = hbarΦ (cbar - 1) (1 - Φinf);
         hbarΦ = hbarΦ /. Solve[ε == (htotbar / zbar), hbarΦ][[1]]
```

Out[7]=
$$\frac{\text{cbar tbar } \epsilon}{(-1 + \text{cbar}) \ (-1 + \epsilon) \ (-1 + \Phi \text{inf})}$$

On page 134 we showed that Φ_∞ was a function of porosity only. We therefore obtain the mean pore height $\bar{h}_\Phi$ as a function of porosity, mean coverage and mean fibre thickness:

In[8]:= **Φinf = 1 + ε (2 + (-3 + ε) ε) / Log[ε];**
      hbarΦ

Out[9]=
$$\frac{\text{cbar tbar Log}[\epsilon]}{(-1 + \text{cbar}) \ (-1 + \epsilon) \ (2 + (-3 + \epsilon) \ \epsilon)}$$

When we plot $\bar{h}_\Phi$ and $\bar{h}$ against porosity we see observe that $\bar{h}_\Phi > \bar{h}$ as expected:

In[9]:= **cbar = 10;**
      LogPlot[{hbar / tbar, hbarΦ / tbar}, {ε, 0.01, 1},
        PlotStyle → {Dashed, {}}, AxesLabel → {"ε", "h̄"}]
      cbar =.

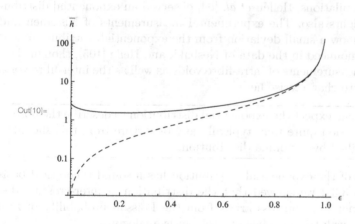

We note also that Niskanen and Rajatora's expression gives $\bar{h} \to 0$ as $\epsilon \to 0$, as expected. When pores of height zero are excluded however, $\bar{h}_\Phi \to \infty$ as $\epsilon \to 0$. This behaviour arises because at low porosities, and hence high fractional contact areas, the expected number of pore heights decreases below 1 and thus the mean pore height must increase to compensate for this unlikely structural characteristic. It is reasonable to assume however that the mean pore height will increase with porosity, so a sensible lower limit of porosity is obtained by determining the location of the minimum in our plot of $\bar{h}_\Phi$ against porosity.

```
In[12]:= dhbarΦ = Simplify[D[hbarΦ / tbar, ε]];
        FindRoot[dhbarΦ (cbar - 1) / cbar, {ε, .2}]
        Φinf /. %
```

```
Out[13]= {ε → 0.205589}
```

```
Out[14]= 0.814734
```

So we may consider our expression to be valid for porosities greater than about 0.2 and hence fractional contact areas less than about 0.8 and thus suitable to describe most fibrous networks where pore size is likely to be an issue. The fact that the expression we derived for $\bar{h}_\Phi$ is less sensitive to porosity than that of Niskanen and Rajatora [109] indicates that as porosity decreases, as in a densifying process such as pressing or calendering, we close pores and generate contacts as well as reducing pore height.

An alternative theoretical analysis of the pore height distribution is provided by Dodson [32] who used the binomial distribution to obtain a probability function for pores with discrete heights given by an integer number of fibre thicknesses. The resulting distribution resembles an exponential distribution and was subsequently shown to agree well with measurements made on paper samples with differing degrees of in-plane uniformity [66]. In their fibre network simulations, Hellén et al. [64] observed an exponential distribution of pore heights also. The experimental measurements of Niskanen and Rajatora [109] show a small deviation from the exponential distribution, and this is more pronounced in the data of Nesbakk and Helle [105], though their data included measurements of intra-fibre voids as well as the inter-fibre voids we have sought to characterise here.

> As a rule, we can expect the exponential distribution to describe the pore height distribution quite well; typically any departure from this should be well described by a gamma distribution.

In fact, the use of the exponential distribution has a sound theoretical basis also: in Section 3.3.2 we showed that the distribution of inter-crossing distances in a planar fibre process arises from the Poisson probability of zero events in an interval; the pore height problem is analogous.

5.4 Porous Anisotropy

Of course, the pore heights that we have just considered characterise the same voids for which we have obtained in-plane dimensions. Collecting our expressions for the mean in-plane pore diameter and the mean pore height we have:

$$\bar{d} = \frac{2}{\log(1/\epsilon)} \, \bar{\omega} \qquad (5.10)$$

$$\bar{h} = \frac{\bar{c}}{\bar{c} - 1} \frac{\log(1/\epsilon)}{(1 - \epsilon) \, (2 - (3 - \epsilon) \, \epsilon)} \, \bar{t} \, . \qquad (5.11)$$

We observe that the mean in-plane pore diameter is proportional to the mean fibre width and the mean pore height is proportional to the mean fibre thickness. For networks of sufficiently high coverage, the first term in Equation 5.11 is redundant so the characteristic pore dimensions depend only on network porosity for fibres with given height and width. The nature of this dependence is revealed by plotting pore height in units of fibre thickness against in-plane pore diameter in units of fibre width:

In[1]:= **hbar** =
    cbar tbar Log[1 / ε] / ((cbar - 1) (1 - ε) (2 - (3 - ε) ε));
    dbar = 2 ωbar / Log[1/ε];
    ParametricPlot[{hbar / tbar, dbar / ωbar} /. cbar → 10⁶,
        {ε, .21, .9}, AspectRatio → 1 / GoldenRatio,
        AxesLabel → {"h̄/t̄", "d̄/ω̄"}, PlotRange → All]

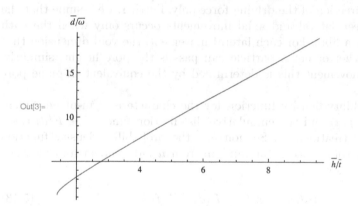

Out[3]=

The plot shows an almost linear relationship between our perpendicular measures of pore dimensions with $\bar{h}/\bar{t}$ always less than $\bar{d}/\bar{\omega}$. The maximum value of $\bar{h}/\bar{d}$ is,

In[4]:= **Limit[hbar / dbar, ε → 1]**

Out[4]= $\dfrac{\text{cbar tbar}}{2 \, (-1 + \text{cbar}) \, \text{ωbar}}$

Since ribbon-like or collapsed fibre will lie essentially in the plane, we typically have $\bar{\omega} > \bar{t}$, so $\bar{t}/\bar{\omega} \leq 1$; also, $\bar{c}/(\bar{c} - 1) > 1$ so we may state that always,

$$\bar{h} < \frac{\bar{d}}{2} \; . \tag{5.12}$$

Recent experimental data arising from tomographic analysis of paper confirms this dependence [136].

This anisotropy of pore dimensions has important consequences for the use of layered stochastic fibrous materials as filters. In Section 5.2 we derived a multi-planar model for the distribution of the smallest pore dimension encountered in a path from one side of the network to the other. Given that the mean pore height is always less than the mean in-plane pore dimension it is likely that the narrowest dimensions in a path from one side of the network to the other will often be associated with a pore height rather than an in-plane pore dimension. To investigate this further, we proceed to derive a multi-planar model for the distribution of the smallest dimension in a path from one side of the network to the other following Sampson and Urquhart [146].

Consider a path from one side of the plane of a fibre network to the other, such as that followed by a fluid or particle passing through the network under some driving force acting perpendicular to its plane. We assume that this path consists of alternate axial and lateral movements, axial movements occurring in the direction of the driving force only. Further, we assume that the transition between lateral and axial movements occurs only when the path is obstructed by a fibre. For each lateral movement, the void dimension that determines whether or not a particle can pass is the pore height; similarly, for each axial movement this is determined by the equivalent in-plane pore diameter.

If the probability density function for the characteristic in-plane dimension of voids is $f(d)$ and the cumulative distribution function is $F(d)$ then, from our earlier treatment in Section 5.2, the probability density function for the smallest voids selected at random from n independent and identical distributions of voids is

$$f(d, n) = n \left(1 - F(d)\right)^{n-1} f(d) \; , \tag{5.13}$$

with cumulative distribution function

$$F(d, n) = \int_0^d f(d, n) \, \mathrm{d}d \; . \tag{5.14}$$

Similarly, and denoting the probability density of pore heights $g(h)$ and the cumulative distribution function $G(h)$, the probability density function for the smallest pore heights selected at random from n independent and identical distributions of pore heights is

$$g(h, n) = n \left(1 - G(h)\right)^{n-1} g(h) \; , \tag{5.15}$$

with cumulative distribution function

$$G(h, n) = \int_0^h g(h, n)\, dh \; . \tag{5.16}$$

Since any path from one side of the network to the other always begins and ends with an out-of-plane movement, then a path that consists of n out-of-plane steps will consist of $(n-1)$ in-plane steps. We now unify the notation such that parameter d represents the characteristic dimension of the smallest of $(n-1)$ pore heights and n in-plane pore diameters. This parameter can be thought of as representing the size of the largest particle that could pass from one side of the network to another and hence is that which we seek to measure, or infer, using porometry techniques; it has probability density

$$p(d) = (1 - G(d, n-1))\, f(d, n) + (1 - F(d, n))\, g(d, n-1) \; . \tag{5.17}$$

Equation 5.17 holds for any pair of distributions $f(d)$ and $g(d)$; for the near-planar stochastic fibrous materials of interest here $f(d)$ and $g(d)$ are given by the gamma and exponential distributions, respectively, allowing the probability density $p(d)$ for given n to be stated explicitly in terms of the parameters characterising the mean and variance of those distributions. So for in-plane pores we have:

```
In[5]:= Clear[hbar]
        fd = PowerExpand[PDF[GammaDistribution[α, β / α], d]];
        Fd =
          FunctionExpand[CDF[GammaDistribution[α, β / α], d]];
        fdn[n_] := n (1 - Fd)^(n-1) fd
        Fdn[n_] := 1 - (1 - Fd)^n
```

and for pore heights we have:

```
In[10]:= gd = PDF[ExponentialDistribution[1 / hbar], d];
         Gd = Refine[
           CDF[ExponentialDistribution[1 / hbar], d], d > 0];
         gdn[n_] := n (1 - Gd)^(n-1) gd
         Gdn[n_] := 1 - (1 - Gd)^n
```

So the probability density in Equation 5.17 is given by

```
In[14]:= pd =
         Simplify[(1 - Gdn[n - 1]) fdn[n] + (1 - Fdn[n]) gdn[n - 1]]
```

$$\text{Out[14]}= \; e^{d/hbar} \left(e^{-\frac{d}{hbar}} \right)^n \left[\frac{-1 + n}{hbar} + \frac{d^{-1+\alpha}\, e^{-\frac{d\alpha}{\beta}}\, n\, \alpha^\alpha\, \beta^{-\alpha}}{\mathrm{Gamma}\left[\alpha, \frac{d\alpha}{\beta}\right]} \right] \left(\frac{\mathrm{Gamma}\left[\alpha, \frac{d\alpha}{\beta}\right]}{\mathrm{Gamma}[\alpha]} \right)^n$$

Before considering this expression further, it is illustrative to examine the probability densities of in-plane pore diameters and pore heights. Recall that the coefficient of variation of in-plane pore dimensions is $1/\sqrt{\alpha} = \sqrt{16 - \pi^2}/\pi$ [20] and the parameter β characterising our gamma distribution is the mean. Note also that the exponential distribution for pore heights is characterised fully by the mean. We input these parameters to *Mathematica*:

```
In[15]:= dbar = 2 ωbar / Log[1 / ε];
         hbar =
           cbar tbar Log[1 / ε] / ((cbar - 1) (1 - ε) (2 - (3 - ε) ε));
         β = dbar;
         α = 1 / cvd²;
         cvd = √(16 - π²) / π;
```

We now compare the distributions of in-plane pore dimensions (solid lines) and pore heights (dashed lines) for networks of infinite coverage formed from fibres with circular cross section such that $\bar{t} = \bar{\omega}$. The first plot is generated for a network with porosity, $\epsilon = 0.8$ and the second for a network with $\epsilon = 0.5$:

```
In[20]:= tbar = ωbar = 1; ε = 0.8;
         Plot[{fd, Limit[gd, cbar → ∞]},
          {d, 0, 30}, PlotStyle → {{}, Dashed},
          AxesLabel → {"d̄/ω̄", "f(d) or g(d)"}, PlotRange → All]
         ε = 0.5;
         Plot[{fd, Limit[gd, cbar → ∞]},
          {d, 0, 10}, PlotStyle → {{}, Dashed},
          AxesLabel → {"d̄/ω̄", "f(d) or g(d)"}, PlotRange → All]
         Clear[tbar, ωbar, ε]
```

Out[21]=

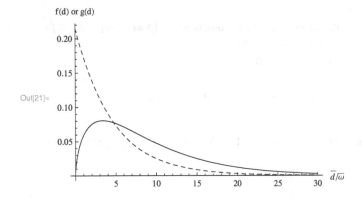

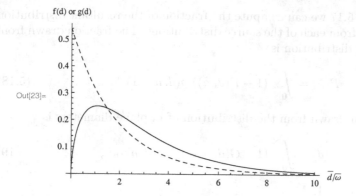

As expected, given our discussion of relative magnitudes of the mean pore height and mean in-plane diameters, we observe that the smallest pore dimensions are more likely to be drawn from the pore height distribution than from in-plane diameter distribution.

To evaluate the probability density for multi-planar structures, we define the number of layers in terms of the porosity in the usual way and must assign a finite value of $\bar{c}$ since this will influence the number of layers n. Our graphic shows the probability density for networks with $\epsilon = 0.5$ (solid line) and $\epsilon = 0.8$ (dashed line):

```
In[25]:= n = cbar / Log[1 / e];
         pd = (1 - Gdn[n - 1]) fdn[n] + (1 - Fdn[n]) gdn[n - 1];
         tbar = ωbar = 1;
         cbar = 10;
         Plot[{pd /. e → .5, pd /. e → .8}, {d, 0, .6},
           PlotRange → All, PlotStyle → {{}, Dashed}]
```

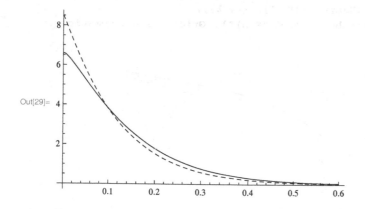

The resultant distributions themselves resemble exponential distributions, with a maximum occurring at small pore diameters at the lower porosity.

From Equation 5.17 we can compute the fraction of the resultant distribution which is drawn from each of the source distributions. The fraction drawn from the pore height distribution is

$$P(h) = \int_0^\infty (1 - F(d, n)) \, g(d, n - 1) \, \mathrm{d}d \,, \qquad (5.18)$$

and the fraction drawn from the distribution of in plane diameters is

$$Q(d) = \int_0^\infty (1 - G(d, n - 1)) \, f(d, n) \, \mathrm{d}d \,, \qquad (5.19)$$

with $P(h) + Q(d) = 1$.

Evaluating Equation 5.18 for networks of mean coverage $\bar{c} = 5$ (dashed line), $\bar{c} = 10$ (solid line) and $\bar{c} = 100$ (dotted line), we see that the fraction of the distribution drawn from the pore height distribution increases with porosity:

```
In[30]:= Clear[cbar]
         Phcbar5 = Table[
             {ε, NIntegrate[((1 - Fdn[n]) gdn[n - 1]) /. cbar → 5,
                 {d, 0, ∞}]}, {ε, .2, .95, .01}];
         Phcbar10 = Table[{ε, NIntegrate[
                 ((1 - Fdn[n]) gdn[n - 1]) /. cbar → 10,
                 {d, 0, ∞}]}, {ε, .2, .95, .01}];
         Phcbar100 = Table[{ε, NIntegrate[
                 ((1 - Fdn[n]) gdn[n - 1]) /. cbar → 100,
                 {d, 0, ∞}]}, {ε, .2, .95, .01}];
         ListLinePlot[{Phcbar10, Phcbar5, Phcbar100},
             PlotStyle → {{}, Dashed, Dotted},
             PlotRange → {{0, 1}, {0, 1}},
             AxesLabel → {"ε", "P(h)"}, GridLines → Automatic]
```

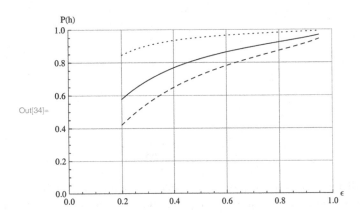

Out[34]=

This fraction increases also with mean coverage since this increases the number of layers and hence the likelihood of encountering a small pore. Note also that the fraction drawn from the pore height distribution is greater than 75 % for porosities greater that 0.65 even in the case where $\bar{c} = 5$.

Of course, the difference between pore height and in-plane dimensions is greater for networks formed from ribbon-like fibres. We can characterise the cross section of these as having aspect ratio $A_{xs} = \bar{\omega}/\bar{t} > 1$. The following code computes the fraction of the distribution drawn from the pore height distribution for networks with porosity $\epsilon = 0.5$ (solid line), $\epsilon = 0.8$ (dashed line) and $\epsilon = 0.9$ (dotted line) formed from fibres with unit width and with $1 \le A_{xs} \le 10$. We observe that as fibres become more ribbon-like, $i.e.$ as A_{xs} increases, an increasing fraction of the distribution is drawn from the pore height distribution:

```
In[35]:= Clear[tbar, ωbar]
         tbar = ωbar / Axs; ωbar = 1; cbar = 10;
         Phe5 = Table[
            {Axs, NIntegrate[((1 - Fdn[n]) gdn[n - 1]) /. ε → 0.5,
             {d, 0, ∞}]}, {Axs, 1, 10, .2}];
         Phe8 = Table[{Axs, NIntegrate[((1 - Fdn[n]) gdn[n - 1]) /.
             ε → 0.8, {d, 0, ∞}]}, {Axs, 1, 10, .2}];
         Phe9 = Table[{Axs, NIntegrate[((1 - Fdn[n]) gdn[n - 1]) /.
             ε → 0.9, {d, 0, ∞}]}, {Axs, 1, 10, .2}];
         ListLinePlot[{Phe5, Phe8, Phe9}, PlotStyle →
            {{}, Dashed, Dotted}, PlotRange → {{0, 10}, {0.5, 1}},
          AxesLabel → {"Axs=ω̄/t̄", "P(h)"}, GridLines → Automatic]
         Clear[tbar, ωbar]
```

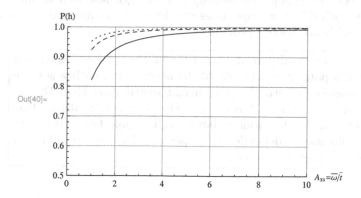

Out[40]=

To investigate the influence of cross-sectional aspect ratio on the mean pore diameter, we must constrain our problem such that the cross-sectional area of the fibres is constant. The simplest way to achieve this is to assume that $\bar{\omega}\,\bar{t}$ is constant. For fibres of unit cross-sectional area and unit width, we have therefore $\bar{t} = 1/\sqrt{A_{\mathrm{xs}}}$.

```
In[42]:= Clear[cbar, ε]
         ωbar = 1; tbar = ωbar/√Axs ; cbar = 10;
         ε = .8;
         dbarAxs = Table[{Axs, NIntegrate[d pd, {d, 0, ∞}]},
             {Axs, 1, 10, .5}];
         ListLinePlot[dbarAxs, PlotRange → All,
           AxesLabel → {"Axs=ω̄/t̄", "d̄/ω̄"}]
```

Out[46]=

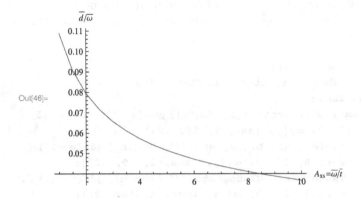

We observe then that the mean pore diameter decreases as the cross-sectional aspect ratio increases, *i.e.* as fibres become more ribbon-like. The use of such fibres in non-woven textiles is well established, see *e.g.* [160] and control of electrospinning techniques is now sufficient to realise fibres with a range of different cross-sectional geometries [88, 90]. Accordingly, selection or manufacture of fibres with different geometries can be used to influence network pore structure and hence barrier properties.

> For networks with typical porosity, *i.e.* $\epsilon > 0.5$, the smallest dimension encountered in a path from one side of the network to another is more likely to be associated with the pore height distribution than the in-plane diameter distribution. Accordingly, the pore height distribution dominates, for example, filtration behaviours. The pore height distribution can be influenced independently of porosity by choosing fibres of different cross-sectional geometry.

5.5 Tortuosity

In our treatment of the pore size distribution for multi-planar networks, we assumed the path through the network to consist of alternate axial and lateral movements. A consequence of such movements is that the length of the path from one side of the network to another is greater than the distance between the two surfaces. We denote the mean path length from one side of the network to the other is $\bar{l}$ and the distance between the surfaces is the mean network thickness $\bar{z}$. The tortuosity factor τ is given by the ratio of these lengths:

$$\tau = \frac{\bar{l}}{\bar{z}} .$$ (5.20)

The distance travelled by fluids in a diffusion type process will depend upon the flow regime, but nonetheless is strongly influenced also by porosity [162]. We therefore consider tortuosity as a structural parameter only and follow the treatment of Foscolo *et al.* [54] for isotropically porous materials.

Consider a fluid element passing through a near-planar fibre network with porosity ϵ under some driving force acting perpendicular to the plane of the network. Fluid will move axially, *i.e.* in the direction of the driving force, until it encounters a fibre whereupon it will move laterally. The probability that a movement is interrupted by encountering a solid is $(1 - \epsilon)$ and the probability that a path continues in a given direction is ϵ. On entering the network, the fraction of the fluid progressing axially a distance δl is ϵ, the remaining fraction $(1 - \epsilon)$ progresses laterally; of this fraction a further fraction ϵ will progress axially at the next transition. It follows that the fraction of fluid that travels a *total* distance $i\,\delta l$ after i transitions is $\epsilon\,(1 - \epsilon)^{i-1}$. The tortuosity of the network is therefore given by,

$$\tau(\epsilon) = \sum_{i=1}^{\infty} i\,\epsilon\,(1 - \epsilon)^{i-1} ,$$ (5.21)

which yields a conveniently simple expression on evaluation:

```
In[1]:= Simplify[Sum[i ϵ (1 - ϵ)^{i-1}, {i, 1, ∞}]]

Out[1]= 1/ϵ
```

This dependence of tortuosity is intuitively reasonable. For more open structures, the deviation in the path length is reduced and in the limit as $\epsilon \to 1$, then so does tortuosity.

There are few direct measures of the tortuosity of random fibre networks reported in the literature. Recent advances in X-ray micro-tomographic techniques have allowed faithful three-dimensional imaging of fibre networks [136–138, 147, 154] and have been used to characterise their pore structure [68]. Using such images as a framework for their computations, Goel *et al.* [58] performed random-walk simulations to determine the tortuosity of paths through the void space of paper samples. These simulations agreed well with data arising from simulations of network conductivity and show a decrease of tortuosity with porosity consistent with our treatment. Interestingly, the data of Goel *et al.* [58] reveal also a dependence of tortuosity on whether the direction of the driving force is in the plane of the network or perpendicular to it. This arises because of the anisotropy of the pore space: paths generated by a driving force perpendicular to the plane encounter fibres within shorter intervals than those generated by a driving force in the plane. Analytic description of the anisotropy of tortuosity is an outstanding theoretical problem, though we might envisage that a random-walk with step lengths drawn alternately from appropriate gamma and exponential distributions may compare favourably with simulation data.

5.6 Distribution of Porosity

We have seen earlier that stochastic fibre materials exhibit a distribution of local average areal densities in the plane. If we assume that the porosity of the network ϵ is independent of the local average areal density, then the relationship between local average areal density $\tilde{\beta}$ and local average network thickness $\tilde{z}$ can be expressed in terms of porosity and the density of the fibres, ρ_f:

$$\tilde{z} = \frac{\tilde{\beta}}{(1 - \epsilon) \rho_f} .$$
(5.22)

It follows that the variance of local thickness in the uniform porosity case is,

$$\sigma^2(\tilde{z}) = \frac{\sigma^2(\tilde{\beta})}{(1 - \epsilon)^2 \rho_f^2} ,$$
(5.23)

and from the central limit theorem, we expect the distribution of local thickness to be Gaussian (normal). Gaussian distributions of thickness and areal density are reported by Dodson *et al.* [37] and by Sung *et al.* [157] for analyses of paper samples using twin-laser profilometry and β-radiography for thickness and areal density measurements, respectively. These data show a strong correlation between local thickness and local areal density and that this correlation persists for networks subjected to a calendering treatment [157]. Similar correlations are reported by Schultz-Eklund *et al.* [152] using a contact measurement technique. An important consequence of this correlation is that there is a distribution of local density and hence local porosity in the plane of the network. We illustrate this using an extruded polymer non-woven filter with mean areal density 40 g m^{-2} as an example; its microstructure is shown in Figure 5.7. The textures plotted in Figure 5.8 represent a region with dimensions 9.9 mm $\times$ 9.1 mm of this filter and show its local average areal density, as obtained by calibrated β-radiography, and its local average thickness, as obtained using twin-laser profilometry. Measurements were made on square inspection regions of side 100 μm; details of the relevant experimental techniques, including details of image registration procedures are provided in [157].

Figure 5.7. Micrograph showing fine structure of the non-woven filter used to obtain the textures plotted in Figure 5.8. (Courtesy D.S. Keller. Reproduced with permission)

The texture on the right of Figure 5.8 represents the distribution of local density for the same region and was obtained by dividing the local average areal density at each location by its local average thickness. Qualitatively, we observe similarities between the textures of the local thickness map and that of local areal density, this being consistent with these two properties being correlated. The texture of the local density map differs from the other two, but remains manifestly non-uniform.

5.6.1 Bivariate Normal Distribution

As the local average grammage and thickness are distributed according to normal distributions and are correlated, we may model their distribution using the bivariate normal distribution which we encountered on page 51. For a network of variable porosity, we have

$$\tilde{z} = \frac{\tilde{\beta}}{(1 - \tilde{\epsilon})\,\rho_f} \, , \tag{5.24}$$

such that

$$(1 - \tilde{\epsilon}) = \frac{\tilde{\beta}}{\tilde{z}\,\rho_f} \, , \tag{5.25}$$

and

$$\sigma_x^2(\tilde{\epsilon}) = \frac{1}{\rho_f^2}\,\sigma_x^2\left(\frac{\tilde{\beta}}{\tilde{z}}\right) \, , \tag{5.26}$$

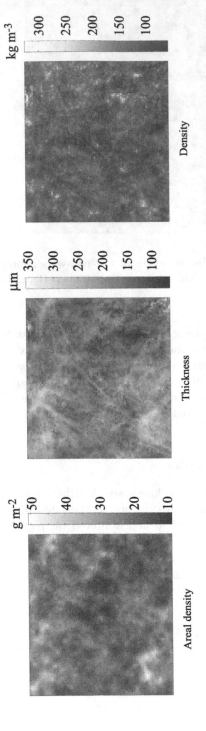

Figure 5.8. Textures showing distributions of local average areal density, thickness and density for an extruded polymer non-woven filter with mean areal density 40 g m^{-2}. Each map represents the same region with dimensions 9.9 mm × 9.1 mm and represents data arising from measurements at the 100-μm scale of inspection. (Courtesy D.S. Keller. Reproduced with permission)

where the subscript x is included since we expect the observed variance to depend upon the scale of inspection x.

If $\tilde{\beta}$ and $\tilde{z}$ are bivariate normally distributed, then we have [39]

$$\sigma_x^2(\tilde{\epsilon}) = \frac{1}{\rho_f^2} \left(\frac{\bar{\beta}}{\bar{z}} \right)^2 \left(\frac{\sigma_x^2(\tilde{\beta})}{\bar{\beta}^2} - \frac{2\,Cov_x(\tilde{\beta},\tilde{z})}{\bar{\beta}\,\bar{z}} + \frac{\sigma_x^2(\tilde{z})}{\bar{z}^2} \right), \qquad (5.27)$$

where $Cov_x(\tilde{\beta},\tilde{z})$ is the covariance of $\tilde{\beta}$ and $\tilde{z}$ at zone size x as given by

$$Cov_x(\tilde{\beta},\tilde{z}) = \overline{\tilde{\beta}\,\tilde{z}} - \bar{\beta}\,\bar{z} \qquad (5.28)$$
$$= \overline{(1-\tilde{\epsilon})\,\rho_f\,\tilde{z}^2} - (1-\bar{\epsilon})\,\rho_f\,\bar{z}^2 \qquad (5.29)$$

Recall from Section 2.5 that the covariance of two random variables can be expressed in terms of their standard deviation and the correlation coefficient ρ_{corr}:

$$Cov_x(\tilde{\beta},\tilde{z}) = \sigma_x(\tilde{\beta})\,\sigma_x(\tilde{z})\,\rho_{corr} . \qquad (5.30)$$

We note also that the mean porosity of the network is given by

$$\bar{\epsilon} = 1 - \frac{\bar{\beta}}{\rho_f\,\bar{z}} , \qquad (5.31)$$

so, the variance of porosity is given by

$$\sigma_x^2(\tilde{\epsilon}) = (1-\bar{\epsilon})^2 \left(cv_x^2(\tilde{\beta}) - 2\,cv_x(\tilde{\beta})\,cv_x(\tilde{z})\rho_{corr} + cv_x^2(\tilde{\beta}) \right), \qquad (5.32)$$

and the coefficient of variation of porosity is given by

$$cv_x(\tilde{\epsilon}) = \left(\frac{1}{\bar{\epsilon}} - 1 \right) \sqrt{cv_x^2(\tilde{\beta}) - 2\,cv_x(\tilde{\beta})\,cv_x(\tilde{z})\rho_{corr} + cv_x^2(\tilde{\beta})} . \qquad (5.33)$$

We observe from Equation 5.33 that the coefficient of variation of porosity decreases with increasing porosity. Surfaces showing the coefficient of variation of porosity as a function of the coefficients of variation of local areal density and thickness, as given by Equation 5.33, are shown in Figure 5.9. These surfaces were computed for the case where $\bar{\epsilon} = 0.5$ such that the term in parentheses in Equation 5.33 takes unit value. We observe that a saddle develops in the surface as ρ_{corr} increases such that in the case of perfect correlation, when $\rho_{corr} = 1$ and $cv_x(\tilde{\beta}) = cv_x(\tilde{z})$, we have $\tilde{\epsilon} = \bar{\epsilon}$ and $cv_x(\tilde{\epsilon}) = 0$. Note also that the data of Oba [112] show $0.5 \leq \rho_{corr} \leq 0.9$ for networks of wood pulp fibres; these data show also that ρ_{corr} is very insensitive to calendering.

At mean porosities greater than 0.5, the coefficient of variation of porosity will be lower than those shown on the vertical axes in Figure 5.9. The figure shows also that for a given coefficient of variation of local areal density $cv_x(\tilde{\beta})$, the coefficient of variation of porosity is lowest when $cv_x(\tilde{\beta}) \approx cv_x(\tilde{z})$. This approximation holds for the low density networks considered by Oba [37, 112] so we may obtain an approximate form of Equation 5.33:

$$cv_x(\tilde{\epsilon}) \approx \left(\frac{1}{\bar{\epsilon}} - 1\right) \sqrt{2(1 - \rho_{\mathrm{corr}})}\, cv_x(\tilde{\beta}) \,. \tag{5.34}$$

We note that a linear relationship between the coefficients of variation of local density and local areal density is reported by Dodson *et al.* [37].

> As a rule-of-thumb, we state that the coefficient of variation of porosity is proportional to that of local average areal density with the constant of proportionality determined by the porosity of the network and the degree of correlation between local average areal density and thickness.

We may readily determine the distribution of local porosities numerically in *Mathematica*. In the example that follows, we consider a network with mean areal density $\bar{\beta} = 60$ g m^{-2} formed from cellulose fibres with density $\rho_f = 1.5$ g cm^{-3} such that the network has mean porosity $\bar{\epsilon} = 0.8$ with coefficients of variation of local areal density and thickness, $cv_x(\bar{\beta}) = cv_x(\bar{z}) = 0.1$ and correlation $\rho_{\mathrm{corr}} = 0.6$. We input these parameters to *Mathematica* taking care to specify consistent units:

```
In[1]:=  βbar = 60; (* g m⁻² *)
         ρf = 1.5; (* g cm⁻³ *)
         ebar = 0.8;
         zbar = βbar / ((1 - ebar) ρf) (* μm *);
         cvz = cvβ = 0.1;
         σz = cvz zbar; σβ = cvβ βbar;
         ρcorr = .6;
```

We generate a millon pairs of bivariate normally distributed $(\tilde{\beta}, \tilde{z})$ using **RandomReal** and calling **MultinormalDistribution** from the package **MultivariateStatistics**:

```
In[8]:=  Needs["MultivariateStatistics`"]
         SeedRandom[1]
         βz = RandomReal[MultinormalDistribution[{βbar, zbar},
             {{σβ², ρcorr σβ σz}, {ρcorr σβ σz, σz²}}], 1 000 000];
```

Mathematica's graphics engine would struggle to plot a million data points, so to generate a plot of local average thickness against local average areal density we select 2,000 points at random from our data using **RandomSample**:

```
In[11]:=  ListPlot[RandomSample[βz, 2000],
          AxesLabel → {"β̃", "z̃"}, PlotRange → All]
```

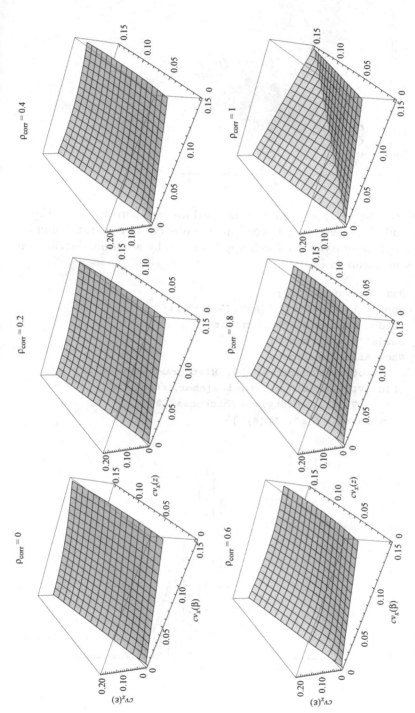

Figure 5.9. Surfaces showing coefficient of variation of porosity as a function of coefficients of variation of areal density and thickness for networks with bivariate normal thickness and areal density with differing correlation. Surfaces computed for $\bar{\epsilon} = 0.5$

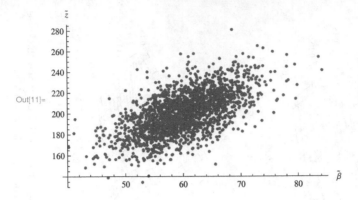

Out[11]=

From each pair of $(\tilde{\beta}, \tilde{z})$ we compute the local average porosity using Equation 5.31 and thus generate a histogram. We observe that the distribution of local average porosity is itself well approximated by a normal distribution with the same mean and variance.

```
In[12]:= βzT = Transpose[βz];
        ϵnet = 1 - (βzT[[1]] / (βzT[[2]] ρf));
        ϵstd = StandardDeviation[ϵnet];
        Needs["Histograms`"]
        Show[Histogram[ϵnet,
           HistogramCategories → 30, HistogramScale → 1],
          Plot[PDF[NormalDistribution[ϵbar, ϵstd], e],
           {e, .7, 1}, PlotStyle → Thickness[.005]],
          AxesLabel → {"ϵ", "f(ϵ)"}]
```

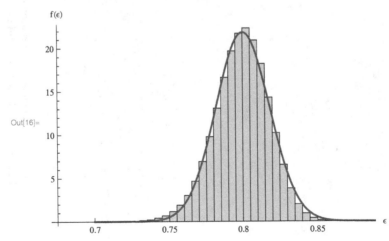

Out[16]=

The distribution does exhibit a weak negative skew however:

In[17]:= **Skewness[ɛnet]**

Out[17]= -0.287457

This is unaffected by mean porosity, but a slightly larger skew is observed for lower correlations. The coefficient of variation of porosity is close to the approximation given by Equation 5.34:

In[18]:= **StandardDeviation[ɛnet] / Mean[ɛnet]**

(1 / ebar - 1) $\sqrt{2\ (1 - \rho \text{corr})}$ cvβ

Out[18]= 0.0228515

Out[19]= 0.0223607

5.6.2 Implications for Network Permeability

A natural consequence of an in-plane distribution of porosity is that we expect this to give rise to a distribution of permeabilities for flow perpendicular to the plane of the network. Variability of local flow rate at small scales is difficult to determine experimentally, though Schweers and Löffler [153] used thermal anemometry to measure the flow rate of air at a distance of 1 mm from the face of a highly porous non-woven filter; they observed a coefficient of variation of local flow velocity of 0.3 at the 0.5-mm scale of inspection and commented that this would probably be greater within the filter.

A classical model for the global average flow rate through a porous material is the Kozeny-Carman equation, as given by

$$v = \frac{1}{K_{kc}} \frac{\Delta P \, d_p^2}{\nu \, \bar{z}} \frac{(1 - \epsilon)^2}{\epsilon^3}. \tag{5.35}$$

where v is the flow velocity $(\mathrm{m\,s^{-1}})$, K_{kc} is the Kozeny constant, ΔP is the pressure drop $(\mathrm{N\,m^{-2}})$, ν is the fluid viscosity $(\mathrm{N\,m^{-2}\,s})$, and d_p is the characteristic dimension (m) of the particles that constitute the porous medium. For many fibrous materials, the Kozeny constant has been determined experimentally to be about 5 [12]; in fact, the term 'constant' is not really appropriate, as the value of K_{kc} is known to depend on the porosity of the network [65]. Nevertheless, Equation 5.35 has proved a useful and reliable tool for many practical purposes and we may use it to provide insights into the influence of network structure on the distribution of local average flow rates. For a recent and extensive review of flow in fibrous materials, see Pan and Zhong [118].

From Equation 5.35, and assuming fluid and flow conditions to be constant, the local average flow rate through a fibre network depends on the local thickness and porosity only:

$$\tilde{v} \propto \frac{1}{\tilde{z}} \frac{(1 - \tilde{\epsilon})^2}{\tilde{\epsilon}^3} . \tag{5.36}$$

Strictly, the Kozeny-Carman equation applies to one-dimensional flows only. We might expect lateral flow distances to be of order one network thickness however, so may assume these to be negligible for the purposes of our analysis. Using the values of $\tilde{z}$ and $\tilde{\epsilon}$ obtained earlier, we may compute the distribution of local flow rates through our network in arbitrary units from Equation 5.36:

In[20]:= **v = (1 / βzT[[2]]) (1 - enet)² / enet³;**
Histogram[v,
**   HistogramCategories → 30, HistogramScale → 1]**

Out[21]=

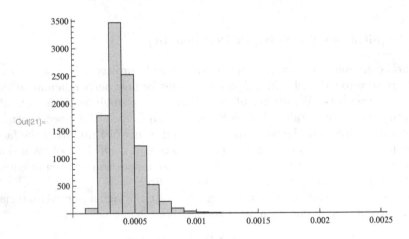

We observe that the distribution has a positive skew and the coefficient of variation of local average flow rates is considerably greater than that of either local porosity or local thickness ($cv(\tilde{\epsilon}) \approx 0.02$, $cv(\tilde{z}) \approx 0.1$):

In[22]:= **StandardDeviation[v] / Mean[v]**

Out[22]= **0.331447**

Collected results for the influence of network uniformity on the local flow velocities are shown in Figures 5.10 and 5.11. Calculations were performed for networks with mean areal density $\bar{\beta} = 60$ g m^{-2} and mean porosity $\bar{\epsilon} = 0.8$ formed from fibres with density $\rho_f = 1.5$ g cm^{-3}. The coefficient of variation

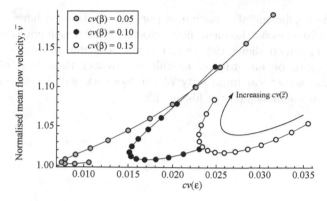

Figure 5.10. Mean flow velocity plotted against coefficient of variation of local porosity for networks with $\bar{\beta} = 60$ g m^{-2}, $\bar{\epsilon} = 0.8$, $\rho_f = 1.5$ g cm^{-3}

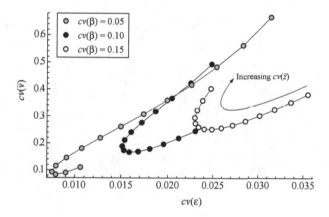

Figure 5.11. Coefficient of variation of local flow velocity plotted against coefficient of variation of local porosity for networks with $\bar{\beta} = 60$ g m^{-2}, $\bar{\epsilon} = 0.8$, $\rho_f = 1.5$ g cm^{-3}

local thickness was varied in the range $0.01 \leq cv(\tilde{z}) \leq 0.15$ and for each value of $cv(\tilde{\beta})$ considered, this influenced the coefficient of variation of local porosity, as plotted on the abscissa; the correlation between local average thickness and areal density was held constant at $\rho_{\text{corr}} = 0.8$. Mean flow rates were normalised by that obtained for a network with uniform porosity and thickness, *i.e.* from Equation 5.36 with $\tilde{\epsilon} = \bar{\epsilon}$ and $\tilde{z} = \bar{z}$.

From Figure 5.10 we observe that the relationship between mean flow rate and the coefficient of variation of porosity is not one-to-one and the mean flow rate $\bar{v}$ initially decreases as $cv(\tilde{z})$ increases due to its influence on $cv(\tilde{\epsilon})$; the subsequent increase is attributable to the increasing influence of the local average thickness $\tilde{z}$ on the local average flow rate. Importantly, even at the low values of the coefficient of variation of porosity considered here, we observe significant influences on the mean flow velocity. Similar behaviour is observed in Figure 5.11 which shows the coefficient of variation of local average flow velocity $cv(\tilde{v})$ to be an order of magnitude greater than that of porosity. Evidently the structural uniformity of the network will be of considerable importance to manufacturers of fibrous filters.

6

Stochastic Departures from Randomness

6.1 Introduction

So far we have considered only random fibre networks where fibre centres are distributed according to a point Poisson process in the plane and fibre axes have a uniform distribution of orientations to any given direction. When discussing the fractional between-zones variance in Section 4.2.3, we remarked that industrially manufactured fibre networks often exhibit a greater variance of local areal density than a random fibre network formed from the same constituent fibres and that this arises as a consequence of fibre interactions during the forming process. A competing effect occurs during the evolution of a fibre network during filtration of a suspension and results as a consequence of the distribution of permeabilities discussed in the previous section: fibres are deposited preferentially in high permeability regions of the evolving network. Since permeability regions with high permeability will typically have low areal density and thickness, the network is 'smoothed' [145] or 'self-heals' [62, 95, 111] during its evolution. Of course, if the first layers of an evolving structure are highly non-uniform as a result of clumping, or fibre *flocculation*, then we expect the contribution of smoothing to be greater. Experimental evidence suggests that this is indeed the case, but that the smoothing is not a sufficiently strong process to overcome the influence of fibre clumping [145].

Outside the laboratory, most commercial web-forming processes are very much directional and result in a continuous web being wound onto a reel at the end of the process. An inevitable consequence of this directionality is that fibres have an increased probability of being aligned close to the direction of manufacture. Detailed discussion of fibre orientation distributions in paper is provided by Niskanen [107]; an overview of fibre orientation in non-woven textiles, where orientation may be induced by *e.g.* carding techniques, including discussion of image analytic techniques for its quantification is provided in the series of articles by Pourdeyhimi *et al.* [126–130]; alternative approaches to the quantification of fibre orientation distribution are provided in, *e.g.* [60, 139, 169].

6.2 Fibre Orientation Distributions

Several analytic distributions have been used to model the fibre orientation distribution in paper and non-woven textiles. We begin by introducing some of the more commonly encountered of these and note that whereas for straight fibres these distributions describe the orientation of the full fibre length to some given direction, for curved fibres they describe the orientation of fibre *segments*. Functions are given here in a form such that the maximum and mean fibre orientation occurs at an angle of $\pi/2$, *i.e.* in the direction of manufacture. Sometimes there is an offset from this angle, and the maximum orientation occurs at an angle $\pm\psi$ to the direction of manufacture, see *e.g.* [60]; in this instance, parameter θ should be replaced by $(\theta \pm \psi)$.

6.2.1 One-parameter Cosine Distribution

The one-parameter cosine distribution was used to model fibre orientation in the early work of Corte and Kallmes [19] following the work of Cox [21] who used a multi-parameter form of the distribution; it has probability density,

$$s(\theta) = \frac{1}{\pi} - \varepsilon_c \cos(2\,\theta) \tag{6.1}$$

where parameter ε_c is termed the *eccentricity* and Equation 6.1 is applicable in the range $0 \leq \varepsilon_c \leq 1/\pi$. When $\varepsilon_c = 0$ we have $s(\theta) = 1/\pi$ such that the probability density of θ is uniform. A plot of $s(\theta)$ compares favourably with experimental data reported in the literature [131]. Here we plot Equation 6.1 for $\varepsilon_c = 0.1$ (dashed line), $\varepsilon_c = 0.2$ (solid line) and $\varepsilon_c = 1/\pi$ (dotted line):

```
In[1]:=  sθ = 1 / π - εc Cos[2 θ];
         Plot[{sθ /. εc → 0.1, sθ /. εc → 0.2, sθ /. εc → 1 / π},
          {θ, 0, π}, PlotRange → {All, {0, .7}},
          AxesLabel → {"θ", "s(θ"},
          PlotStyle → {Dashed, {}, Dotted}]
```

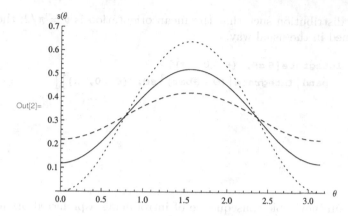

A more intuitive understanding is obtained by rendering a polar plot of the probability density as used by, *e.g.* Niskanen [107]:

```
In[3]:= PolarPlot[{sθ /. εc → 0.1, sθ /. εc → 0.2, sθ /. εc → 1 / π},
        {θ, 0, 2 π}, PlotStyle → {Dashed, {}, Dotted},
        AspectRatio → Automatic]
```

Out[3]=

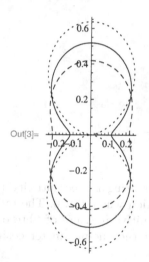

As the distribution is symmetrical, the probability density $s(\theta)$ is defined only on the interval $0 \leq \theta \leq \pi$, so

$$\int_0^\pi s(\theta) \, d\theta = 1 .$$

Our polar plots have been generated for $0 \leq \theta \leq 2\pi$ to reveal the symmetry, so the perimeter of each curve shown is 2. In the case of a uniform fibre orientation distribution, a polar plot of $s(\theta)$ yields a circle of radius $1/\pi$. We have defined the distribution such that the mean orientation is $\bar{\theta} = \pi/2$; the variance is obtained in the usual way:

In[4]:= **θbar = Integrate[θ sθ, {θ, 0, π}]**
       θvar = Expand[Integrate[(θ - θbar)^2 sθ, {θ, 0, π}]]

Out[4]= $\dfrac{\pi}{2}$

Out[5]= $\dfrac{\pi^2}{12} - \dfrac{\pi \, \epsilon c}{2}$

We observe that an inevitable consequence of introducing a preferred orientation to the network is that the variance of the orientation angle decreases with increasing eccentricity:

In[6]:= **Plot[θvar, {ϵc, 0, 1/π}, AxesLabel -> {"ϵ$_c$", "σ^2(θ)"}]**

From the polar plots of $s(\theta)$ we see that increasing the eccentricity results in a narrowing of the distribution in the region where $\theta \approx 0$. The ratio $s(\pi/2)/s(0)$ provides what is perhaps a more intuitive measure of fibre orientation and is termed the *orientation ratio*. For the one-parameter cosine distribution we have,

In[7]:= **ORc = Simplify[(sθ /. θ → (π/2)) / sθ /. θ → 0]**

Out[7]= $\dfrac{1 + \pi \, \epsilon c}{1 - \pi \, \epsilon c}$

6.2.2 von Mises Distribution

The von Mises distribution has probability density,

$$t(\theta) = \frac{e^{-\varepsilon_{vm}\,\cos(2\theta)}}{\pi\,I_0(\varepsilon_{vm})}\,, \tag{6.2}$$

where ε_{vm} is the free parameter controlling the orientation of the distribution and $I_0(\varepsilon_{vm})$ is the zeroth order modified Bessel function of the first kind. As for the one-parameter cosine distribution, when $\varepsilon_{vm} = 0$, Equation 6.2 reduces to a uniform probability density. Here we plot Equation 6.2 for $\varepsilon_{vm} = 1$ (dashed line), $\varepsilon_{vm} = 2$ (solid line) and $\varepsilon_{vm} = 5$ (dotted line):

```
In[8]:= tθ = e^-evm Cos[2 θ] / (π BesselI[0, evm]);
        Plot[{tθ /. evm → 1, tθ /. evm → 2, tθ /. evm → 5},
          {θ, 0, π}, PlotStyle → {Dashed, {}, Dotted},
          AspectRatio → Automatic, PlotRange → All,
          AxesLabel → {"θ", "t(θ)"}]
```

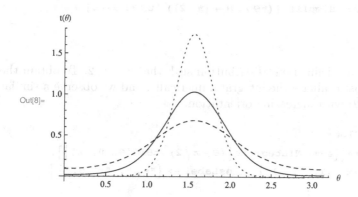

Out[8]=

We remark also that the von Mises distribution converges on the normal distribution as ε_{vm} increases, and note that a normal distribution of fibre orientations is reported by Cheng and Sastry [15]. A polar plot of the von Mises distribution is qualitatively similar to that observed for the one-parameter cosine distribution, though there is no upper bound on ε_{vm} so greater orientation ratios are permitted:

```
In[9]:= PolarPlot[{tθ /. evm → 1, tθ /. evm → 2, tθ /. evm → 3},
          {θ, 0, 2 π}, PlotStyle → {Dashed, {}, Dotted},
          AspectRatio → Automatic, PlotRange → All]
```

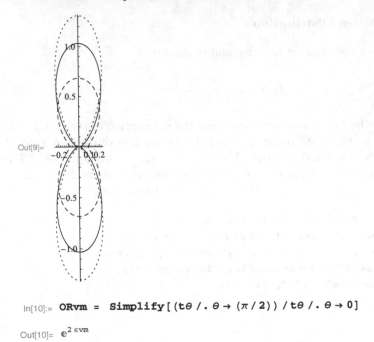

Out[9]=

In[10]:= **ORvm = Simplify[(tθ /. θ → (π/2))/tθ /. θ → 0]**

Out[10]= $e^{2 \epsilon vm}$

Again, we have defined our distribution such that $\bar{\theta} = \pi/2$. To obtain the variance, we must evaluate the integral numerically, and we observe a similar decrease of $\sigma^2(\theta)$ with increasing orientation:

In[11]:= **ListPlot[**
  Table[{evm, NIntegrate[(θ - π/2)² tθ, {θ, 0, π}]},
  {evm, .1, 3, .1}], AxesLabel -> {"ϵ_{vm}", "$\sigma^2(\theta)$"}]

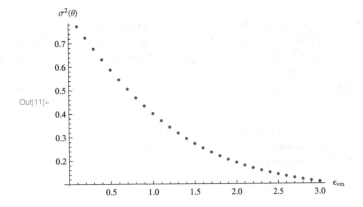

Out[11]=

6.2.3 Wrapped Cauchy Distribution

The wrapped Cauchy distribution, or elliptical distribution, was used by
Schulgasser [151] to describe fibre orientation in paper. It has probability
density,

$$u(\theta) = \frac{1}{\pi} \frac{1 - \varepsilon_{wc}^2}{1 + \varepsilon_{wc}^2 - 2\,\varepsilon_{wc}\,\cos(2\,\theta)} \tag{6.3}$$

with free parameter $-1 \le \varepsilon_{wc} \le 0$. Here we plot Equation 6.3 for $\varepsilon_{wc} = -0.1$
(dashed line), $\varepsilon_{wc} = -0.2$ (solid line) and $\varepsilon_{vm} = -0.3$ (dotted line):

```
In[12]:= uθ = 1 / π (1 - ewc²) / (1 + ewc² - 2 ewc Cos[2 θ]);
         Plot[{uθ /. ewc → (-.1),
           uθ /. ewc → (-.2), uθ /. ewc → (-.3)},
           {θ, 0, π}, PlotStyle → {Dashed, {}, Dotted},
           PlotRange → {All, {0, .6}}, AxesLabel → {"θ", "u(θ)"}]
```

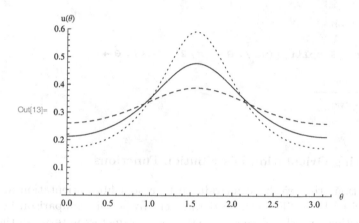

```
In[14]:= PolarPlot[
           {uθ /. ewc → (-.1), uθ /. ewc → (-.2), uθ /. ewc → (-.3)},
           {θ, 0, 2 π}, PlotStyle → {Dashed, {}, Dotted},
           AspectRatio → Automatic,  PlotRange → All,
           AxesLabel → {"θ", "u(θ)"}]
```

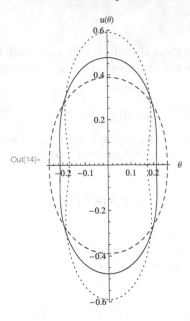

Out[14]=

In[15]:= **ORwc = Simplify[(uθ /. θ → (π / 2)) / uθ /. θ → 0]**

Out[15]= $\dfrac{(-1 + \epsilon wc)^2}{(1 + \epsilon wc)^2}$

6.2.4 Comparing Orientation Distribution Functions

The three functions that we have introduced to model fibre orientation are each parameterised by a different free parameter. To permit comparison between these functions, we therefore require the free parameters in terms of the orientation ratio:

In[16]:= **ɛc = ɛc /. Solve[ORc == OR, ɛc][[1]]**
ɛvm = ɛvm /. Solve[ORvm == OR, ɛvm][[2]]
ɛwc = ɛwc /. Simplify[Solve[ORwc == OR, ɛwc]][[2]]

Out[16]= $\dfrac{-1 + OR}{(1 + OR)\,\pi}$

Out[17]= $\dfrac{Log[OR]}{2}$

Out[18]= $\dfrac{1 - \sqrt{OR}}{1 + \sqrt{OR}}$

Note that for the von Mises and wrapped Cauchy distributions, we take the second of two possible solutions, as the first yields values of ε_{vm} and ε_{wc} outside the applicable range. We now have all three probability densities parameterised in terms of the orientation ratio:

In[19]:= **sθ**
   tθ
   Simplify[uθ]

Out[19]= $\dfrac{1}{\pi} - \dfrac{(-1 + OR)\,Cos[2\,\theta]}{(1 + OR)\,\pi}$

Out[20]= $\dfrac{OR^{-\frac{1}{2}Cos[2\,\theta]}}{\pi\,BesselI\left[0,\ \frac{Log[OR]}{2}\right]}$

Out[21]= $\dfrac{2\,\sqrt{OR}}{\pi\,(1 + OR + (-1 + OR)\,Cos[2\,\theta])}$

In the subsequent plots we use a dashed line for the cosine distribution, a solid line for the von Mises distribution and a dotted line for the wrapped Cauchy distribution. Polar plots of these probability densities for orientation ratios, $OR = 3, 5$, and 7 reveal that they exhibit similar shapes:

```
In[22]:= GraphicsArray[
            {{PolarPlot[Evaluate[{sθ, tθ, uθ} /. OR → 3],
                {θ, 0, 2 π}, PlotStyle → {Dashed, {}, Dotted},
                PlotRange → {{-.25, .25}, {-1, 1}},
                PlotLabel → "OR = 3"],
             PolarPlot[Evaluate[{sθ, tθ, uθ} /. OR → 5],
                {θ, 0, 2 π}, PlotStyle → {Dashed, {}, Dotted},
                PlotRange → {{-.25, .25}, {-1, 1}},
                PlotLabel → "OR = 5"],
             PolarPlot[Evaluate[{sθ, tθ, uθ} /. OR → 7],
                {θ, 0, 2 π}, PlotStyle → {Dashed, {}, Dotted},
                PlotRange → {{-.25, .25}, {-1, 1}},
                PlotLabel → "OR = 7"]}}]
```

Out[22]=

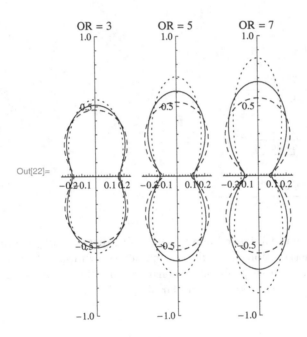

Comparison of the cumulative distribution functions, which must be obtained numerically for the von Mises and wrapped Cauchy distributions, confirms this similarity.

```
In[23]:= Sθ = Integrate[sθ, {θ, 0, θ}];
        OR = 5;
        Show[
         ListLinePlot[Table[{x, NIntegrate[uθ, {θ, 0, x}]},
          {x, 0, π, π/50}], PlotStyle → Dotted],
         ListLinePlot[Table[{x, NIntegrate[tθ, {θ, 0, x}]},
          {x, 0, π, π/50}]],
         Plot[Sθ, {θ, 0, π}, PlotStyle → Dashed],
         AxesLabel → {"θ", "CDF(θ)"}]
        Clear[OR]
```

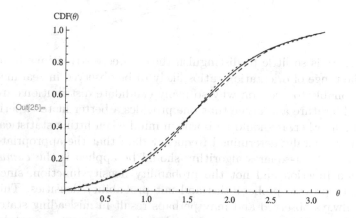

Out[25]=

Naturally, given such similar cumulative distribution functions for our three distributions, we observe that the variance of fibre orientation is similar also for each distribution:

```
In[27]:= Show[ListLinePlot[
         {Table[{OR, NIntegrate[(θ - π/2)² tθ, {θ, 0, π}]},
           {OR, 1, 7, .2}],
          Table[{OR, NIntegrate[(θ - π/2)² uθ, {θ, 0, π}]},
           {OR, 1, 7, .2}]}, PlotStyle → {Dotted, {}}],
         Plot[θvar, {OR, 1, 7}, PlotStyle → Dashed],
         AxesLabel → {"OR", "σ²(θ)"}]
```

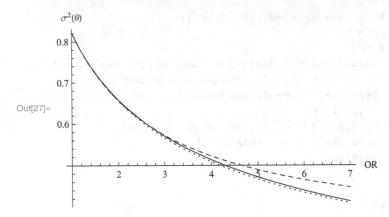

Given that there is so little to distinguish these three distributions from each other in the range of orientation ratios likely to be observed in real materials, it is reasonable to question why so many candidate distributions are proposed in the literature and argue that one provides a better fit to experimental data than another. It should be borne in mind when fitting statistical functions to experimentally determined frequency data that the appropriate function to which a least-squares algorithm should be applied is the *cumulative* distribution function and not the probability density function, since experimental data provide probabilities and not probability densities. This practice is not always observed and may perhaps result in misleading statements as to which probability density function describes data most accurately. It is, nonetheless, convenient that the cumulative distribution functions considered agree so closely over the range of interest. Given that the von Mises and wrapped Cauchy distributions are more likely to require numerical analysis than the cosine distribution and the convenient simplicity of form of the latter, we carry out our subsequent analysis of the influence of fibre orientation on other structural properties of interest using the one-parameter cosine distribution only.

6.2.5 Fibre Crossings

In Section 4.2.1 we derived the expected number of crossings per unit area and the expected number of crossings per fibre for random fibre networks with a uniform distribution of fibre orientations. Here we repeat this analysis using the one-parameter cosine distribution to describe the orientation of fibres in the network following Corte and Kallmes [19]. Recall that the probability that a pair of lines of length λ with centres within an area x^2 and oriented at angle ϕ to each other will intersect is

$$P_{\text{cross},\phi} = \frac{\lambda^2}{x^2} \sin(\phi) \ . \tag{6.4}$$

The probability that *any* pair of lines, with arbitrary orientation, intersect is given by,

$$\int P_{\text{cross},\phi}\, g(\phi)\, \mathrm{d}\phi \ , \tag{6.5}$$

where $g(\phi)$ is the probability density function for ϕ and in Section 4.2.1 we used a uniform distribution for this. In developing theory for oriented networks, we must consider this function further.

If the probability density of fibre orientation is given by the one-parameter cosine distribution, then we have,

In[1]:= **fθ = 1 / π - ϵc Cos [2 θ] ;**

We seek the distribution of the angles ϕ generated by the crossing of a pair of fibres with independent and identical probability densities of fibre orientation θ. Each crossing generates two angles,

$$\phi = |\theta_1 - \theta_2|$$

and

$$\phi = \pi - |\theta_1 - \theta_2| \ .$$

Angle ϕ therefore has probability density

$$g(\phi) = \frac{f(\theta_1 - \theta_2) + f(\theta_2 - \theta_1) + f(\pi - (\theta_1 - \theta_2)) + f(\pi - (\theta_2 - \theta_1))}{2} \ . \tag{6.6}$$

where,

$$f(\theta_1 - \theta_2) = \int_{\phi}^{\pi} f(\theta_1)\, f(\theta_1 - \phi)\, \mathrm{d}\theta_1 \tag{6.7}$$

$$f(\theta_2 - \theta_1) = \int_{0}^{\pi - \phi} f(\theta_1)\, f(\theta_1 + \phi)\, \mathrm{d}\theta_1 \tag{6.8}$$

$$f(\pi - (\theta_1 - \theta_2)) = \int_{0}^{\phi} f(\theta_1)\, f(\pi + (\theta_1 - \phi))\, \mathrm{d}\theta_1 \tag{6.9}$$

$$f(\pi - (\theta_2 - \theta_1)) = \int_{\pi - \phi}^{\pi} f(\theta_1)\, f(\pi + (\theta_1 + \phi))\, \mathrm{d}\theta_1 \ . \tag{6.10}$$

Inputting these integrals to *Mathematica* and simplifying yields the probability density $g(\phi)$:

In[2]:= **fφ1 = FullSimplify[Integrate[**
            fθ (fθ /. θ → (θ - φ)), {θ, φ, π}], 0 ≤ φ ≤ π];
        fφ2 = FullSimplify[Integrate[fθ (fθ /. θ → (θ + φ)),
            {θ, 0, π - φ}], 0 ≤ φ ≤ π];
        fφ3 = FullSimplify[Integrate[fθ (fθ /. θ → (π + (θ - φ))),
            {θ, 0, φ}], 0 ≤ φ ≤ π];
        fφ4 = FullSimplify[Integrate[fθ (fθ /. θ → (π + (θ + φ))),
            {θ, π - φ, π}], 0 ≤ φ ≤ π];
        gφ = Simplify[(fφ1 + fφ2 + fφ3 + fφ4) / 2]
        TrueQ[Integrate[gφ, {φ, 0, π}] == 1]

Out[6]= $\dfrac{1}{\pi} + \dfrac{1}{2} \, \pi \, \epsilon c^2 \, Cos[2 \, \phi]$

Out[7]= **True**

Now, $g(\phi)$ gives the probability density of crossing angles ϕ assuming that all pairs of lines make a crossing. The probability that a given pair of fibres with crossing angle ϕ does in fact generate a crossing is given by Equation 6.4, so the probability of crossings between any pair of fibres is:

In[8]:= **pcrossφ = λ² Sin[φ] / x²;**
        pcross = Expand[Integrate[gφ pcrossφ, {φ, 0, π}]]

Out[9]= $\dfrac{2 \, \lambda^2}{\pi \, x^2} - \dfrac{\pi \, \epsilon c^2 \, \lambda^2}{3 \, x^2}$

such that when $\varepsilon_c = 0$ we recover the result derived on page 111 for networks with uniform orientation. In that derivation we assumed $g(\phi)$ to have uniform probability density following Kallmes and Corte [74]; from the treatment given here we are able to justify this assumption.

The probability density of crossing angles for those fibres which generate a crossing is,

In[10]:= **hφ = Simplify[(gφ pcrossφ) / pcross]**
         TrueQ[Integrate[hφ, {φ, 0, π}] == 1]

Out[10]= $-\dfrac{3 \left(2 + \pi^2 \, \epsilon c^2 \, Cos[2 \, \phi]\right) \, Sin[\phi]}{2 \left(-6 + \pi^2 \, \epsilon c^2\right)}$

Out[11]= **True**

Such that when $\varepsilon_c = 0$ we have the distribution of crossing angles given by a sine distribution:

In[12]:= **hϕ /. ϵc → 0**

Out[12]= $\dfrac{\text{Sin}[\phi]}{2}$

As ε_c increases, we observe an increased probability of crossing angles at small and large ϕ and a decrease in the occurrence of crossings with ϕ close to $\pi/2$, such that for highly oriented networks, the distribution is bimodal:

In[13]:= **Plot [{hϕ /. ϵc → 0, hϕ /. ϵc → .15, hϕ /. ϵc → 1 / π},**
{ϕ, 0, π}, PlotStyle → {{}, Dashed, Dotted},
AxesLabel → {"ϕ", "h(ϕ)"}]

Out[13]=

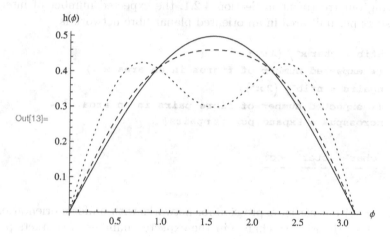

We observe also that, as expected, the mean, $\bar{\phi}$ is not influenced by orientation, but the variance of crossing angles increases:

In[14]:= **ϕbar = Integrate[ϕ hϕ, {ϕ, 0, π}]**
ϕvar = Integrate$\left[$ (ϕ - ϕbar)^2 hϕ, {ϕ, 0, π}$\right]$
Plot$\left[$$\phi$var, {$\epsilon$c, 0, 1 / π}, AxesLabel → $\left\{$"ϵ_c", "σ^2 (ϕ)"$\right\}$$\right]$

Out[14]= $\dfrac{\pi}{2}$

Out[15]= $-\dfrac{26}{9} + \dfrac{\pi^2}{4} + \dfrac{16}{18 - 3\pi^2 \epsilon c^2}$

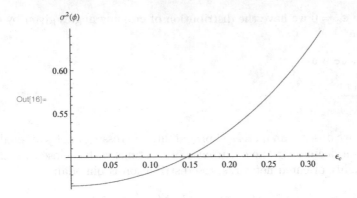

The implication of this result for the distribution of crossing areas is discussed in the next section.

Following our treatment in Section 4.2.1, the expected number of inter-fibre crossings per unit area in an oriented planar fibre network is

In[17]:= **nfib = cbar x² / (λ ω);**
(* expected number of fibres in an area x²*)
npairs = nfib² / (2 x²);
(* expected number of fibre pairs in an area x² *)
ncrosspua = Expand[pcross npairs]

Out[18]= $\dfrac{\text{cbar}^2}{\pi\,\omega^2} - \dfrac{\text{cbar}^2\,\pi\,\epsilon c^2}{6\,\omega^2}$

Again we recover our earlier expression for networks with uniform orientation when $\varepsilon_c = 0$, so the relative change in the expected number of contacts per unit area arising from fibre orientation is

In[19]:= **Simplify[pcross / (pcross /. ec → 0)]**

Out[19]= $1 - \dfrac{\pi^2\,\epsilon c^2}{6}$

We observe that the net influence of fibre orientation on the expected number of contacts per unit area, and hence the expected number of contacts per fibre is rather small:

In[20]:= **Plot[%, {ec, 0, 1/π}]**

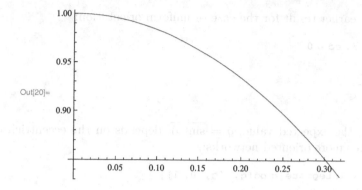

For completeness, the expected number of contacts per fibre is given by

In[21]:= **Expand$\left[$2 ncrosspua x^2 / nfib$\right]$**

Out[21]= $\dfrac{2 \text{ cbar } \lambda}{\pi \, \omega} - \dfrac{\text{cbar } \pi \, \epsilon \text{c}^2 \, \lambda}{3 \, \omega}$

6.2.6 Crossing Area Distribution

A direct consequence of the distribution of fibre orientations $f(\theta)$ influencing the distribution of fibre crossing angles $h(\phi)$ is that the distribution of fibre crossing areas also will be influenced. For fibres of width ω, the expected area of a crossing is

$$a = \frac{\omega^2}{\sin(\phi)} \, . \tag{6.11}$$

To obtain $\overline{\sin(\phi)}$ we must first derive the probability density of the random variable $q = \sin(\phi)$. Following the treatment given on page 146 and using the probability density for crossing angles $h(\phi)$ derived in the previous section we have:

In[1]:= **hϕ = 3 $\left(2 + \pi^2 \, \epsilon \text{c}^2 \, \text{Cos}\left[2\,\phi\right]\right)$ Sin$\left[\phi\right]$ / $\left(2 \left(6 - \pi^2 \, \epsilon \text{c}^2\right)\right)$;**
      ϕ = ArcSin$\left[\text{q}\right]$;
      pdfq = Simplify$\left[$FunctionExpand$\left[2 \, \text{h}\phi \, \text{D}\left[\phi, \, \text{q}\right]\right]\right]$
      TrueQ$\left[$Integrate$\left[\text{pdfq}, \, \{\text{q}, \, 0, \, 1\}\right]\right]$ == 1

Out[3]= $\dfrac{3 \, \text{q} \left(-2 + \pi^2 \left(-1 + 2 \, \text{q}^2\right) \epsilon \text{c}^2\right)}{\sqrt{1 - \text{q}^2} \left(-6 + \pi^2 \, \epsilon \text{c}^2\right)}$

Out[4]= **True**

We recover our earlier result for the case of uniform orientation:

In[5]:= **pdfq /. ϵc → 0**

Out[5]= $\dfrac{q}{\sqrt{1-q^2}}$

As anticipated, the expected value, $\bar{q} = \overline{\sin(\phi)}$, depends on the eccentricity and decreases for more oriented networks:

In[6]:= **qbar = Integrate[q pdfq, {q, 0, 1}]**
       qbar /. ϵc → 0

Out[6]= $\dfrac{3}{8}\,\pi\left(1+\dfrac{2}{-6+\pi^2\,\epsilon c^2}\right)$

Out[7]= $\dfrac{\pi}{4}$

In[8]:= **Plot$\left[$qbar, {ϵc, 0, 1 / π}, AxesLabel → $\left\{$"ϵc", "$\overline{\text{Sin}\,(\phi)}$"$\right\}\right]$**

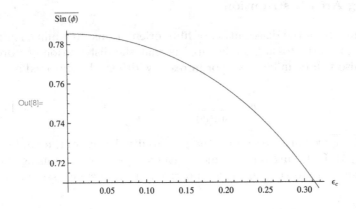

The variance of $\sin(\phi)$, and hence the coefficient of variation, are only weakly influenced by orientation:

In[9]:= **qvar = Integrate$\left[$(q - qbar)^2 pdfq, {q, 0, 1}$\right]$**
       Plot$\left[$qvar, {ϵc, 0, 1 / π}, AxesLabel → $\left\{$"ϵc", "σ^2(q)"$\right\}\right]$
       Plot$\left[\sqrt{\text{qvar}}\,\big/\,\text{qbar}$, {ϵc, 0, 1 / π},
       AxesLabel → {"ϵc", "cv(q)"}$\right]$

Out[9]= $$\dfrac{3\left(-\frac{4}{3}+\frac{1}{320}\,\pi^2\left(30+\left(128-15\,\pi^2\right)\,\epsilon c^2-\dfrac{60}{-6+\pi^2\,\epsilon c^2}\right)\right)}{-6+\pi^2\,\epsilon c^2}$$

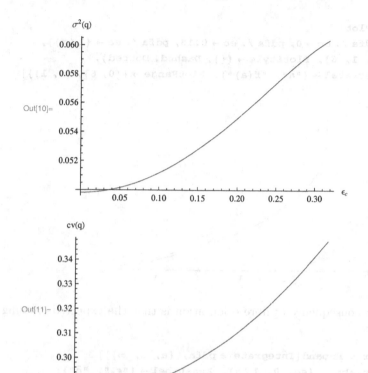

The distribution of crossing areas is given in units of the fibre width squared by the random variable $a = 1/q$ and we determine its probability density as:

In[12]:= ```
q = 1 / a;
pdfa = Simplify[-D[q, a] pdfq, a > 1]
```

Out[13]= $$\dfrac{3\left(-2+\left(-1+\frac{2}{a^2}\right)\pi^2\,\epsilon c^2\right)}{a^2\sqrt{-1+a^2}\,\left(-6+\pi^2\,\epsilon c^2\right)}$$

The resultant probability density is defined in the range $1 \le a \le \infty$. A plot of this probability density for uniform orientation (solid line), $\varepsilon_c = 0.15$ (dashed
```

line) and the most oriented case permitted by the theory, *i.e.* $\varepsilon_c = 1/\pi$ (dotted line) reveals the increased incidence of large crossing areas in more oriented networks:

In[14]:= **LogPlot [**
{pdfa /. ϵc → 0, pdfa /. ϵc → 0.15, pdfa /. ϵc → (1 / π)},
{a, 1, 6}, PlotStyle → {{}, Dashed, Dotted},
AxesLabel → {"a", "f(a)"}, PlotRange → {{0, 6}, {0, 1}}]

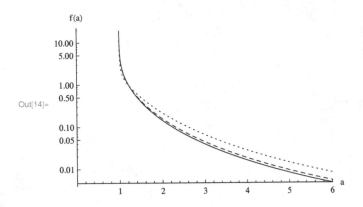

Accordingly, a consequence of fibre orientation is that the expected crossing area increases:

In[15]:= **abar = Expand[Integrate[a pdfa, {a, 1, ∞}]]**
Plot[abar, {ϵc, 0, 1 / π}, AxesLabel → {"ϵc", "ā"}]

Out[15]= $-\dfrac{3\,\pi}{-6 + \pi^2\,\epsilon c^2}$

We estimate the total area of crossings by the product of the mean area of a single crossing and the expected number of crossings per fibre, as derived in the previous section:

In[17]:= **ncpf = 2 cbar λ / (π ω) $\left(1 - \pi^2\ \epsilon c^2 / 6\right)$;**
      atotal = Simplify$\left[$ncpf abar $\omega^2\right]$

Out[18]= cbar $\lambda\ \omega$

> The resultant expression does not include the eccentricity of the network as a variable, so fibre orientation reduces the number of contacts and increases their area with the net result being that the total area of contacts, and hence the fractional contact area of the network, is unaffected.

From their Monte Carlo simulation, Kallmes and Bernier [77] note that increased fibre orientation introduces a similar orientation distribution to Fractional Contact Area, *i.e.* the amount of inter-fibre contact observed along scan-lines was influenced by the direction of scanning. No effect was observed for the absolute contact states; this is perhaps unsurprising since these expressions are functions of the mean coverage only. We expect then that any influence of fibre orientation on mechanical properties arises solely as a consequence of the structural mechanics of the fibres themselves and not from any influence on bonding at regions of inter-fibre contact.

Given the influence of orientation on the number of contacts per fibre, there will be a corresponding influence on the mean inter-crossing distance. Following our derivation of Equation 5.2 for networks with uniform orientation, the mean inter-fibre crossing distance is approximated by:

In[19]:= **gbar = Expand[λ / ncpf - ω / qbar]**

$$\text{Out[19]=}\quad \frac{\pi\ \omega}{2\ \text{cbar}\ \left(1 - \frac{\pi^2\ \epsilon c^2}{6}\right)} - \frac{8\ \omega}{3\ \pi\ \left(1 + \frac{2}{-6 + \pi^2\ \epsilon c^2}\right)}$$

Plotting the mean inter-crossing distance against coverage for uniformly oriented networks (solid line) and highly oriented networks (dashed line) reveals only a weak dependence of mean inter-crossing distance on orientation for planar networks and this is consistent with the simulation data of Kallmes and Bernier [77]:

In[20]:= **Plot[{gbar /. ω → 1 /. ϵc → 0, gbar /. ω → 1 /. ϵc → 0.3},**
      {cbar, 0, 1}, PlotStyle → {{}, Dashed},
      AxesLabel → {"c̄", "ḡ/ω"}]

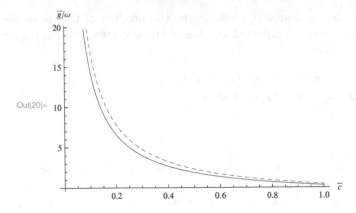

It is reasonable to expect the same weak dependence to persist in multi-planar structures also and we note that if the mean inter-crossing distance is insensitive to orientation, then the mean in-plane pore dimensions will be similarly insensitive, *cf.* Section 5.2. This is confirmed by the data of Castro and Ostoja-Starzewski [13] who report from their simulations that the number of polygons per unit area and the area-frequency of inscribed circle radii is insensitive to the fibre orientation distribution. An additional contributory factor to this insensitivity is that any influence of fibre orientation on the correlation between adjacent polygon sides will be overwhelmed by the inherent clustering of a point Poisson process for fibre centres [40].

6.2.7 Mass Distribution

Schaffnit and Dodson [149] derived the autocorrelation function, and hence the fractional between-zones variance for fibres with centres distributed according to a point Poisson process in the plane and with fibre orientation distributions given by the one-parameter cosine distribution, the wrapped Cauchy distribution and a rectangular distribution, thus extending the treatment considered in Section 4.2.3. The resulting expressions and computations showed the distribution of mass density to be insensitive to fibre orientation for the one-parameter cosine distribution, to be negligibly less uniform with increasing orientation for the wrapped Cauchy distribution and to be slightly more uniform with increasing orientation for the rectangular distribution. In conclusion, Schaffnit and Dodson [149] considered the distribution of mass density to be only slightly influenced, if at all, by the degree of fibre orientation. The finding is consistent with those obtained from experimental studies by Waterhouse [167] and from simulation data by Cresson [23]. An earlier result of Dodson and Fekih [34] suggested that increased orientation would increase the coefficient of variation of local areal density; this was identified as being attributable to an approximation in the derivation.

In a simulation study, Soszyński [155] computed the fractional between-zones variance for networks of rectangular fibres with randomly located fibre centres and with axes oriented with probability density given by a two-parameter cosine distribution with orientation ratio 5,

$$s(\theta) = \frac{1}{\pi} + \frac{\cos(2\,\theta)}{\pi} + \frac{\cos(4\,\theta)}{2\,\pi} \ . \tag{6.12}$$

Soszyński found that for square inspection zones, the fractional between-zones variance was insensitive to the orientation of fibres, in agreement with Schaffnit and Dodson [149]. For rectangular inspection zones however, the fractional between-zones variance was dependent on fibre orientation. Similar anisotropy of the autocorrelation function is reported by Johansson and Hössjer [71] who used a shot-noise approach and simulation for networks with randomly located fibre centres and with axes oriented with probability density given by the von Mises distribution. They observed a small decrease in the autocorrelation, and hence in the fractional-between zones variance, in the range of orientation ratios typical of industrially manufactured networks.

6.3 Fibre Clumping and Dispersion

Here we consider departures from randomness arising due to fibre clumping, or flocculation and smoothing or self-healing. Given that clumping overwhelms any smoothing effect [145] we consider only networks with greater distribution of local coverage and areal density than a random network formed from the same fibres, as is likely to be observed in a finished product. Theory for disperse networks is not well covered in the literature, though some results are provided by Kallmes and Bernier [77] who observed no influence of clumping or dispersion on the fractional contact area and absolute contact states of thin fibre networks.

Perhaps the most complete description of the texture of the mass distribution in near-planar stochastic fibrous materials is provided by Farnood *et al.* [50]. They modelled clusters of fibres as projecting their mass over a disc of diameter $D_d = 2\,R$.

We determine the point autocorrelation function $\alpha(r)$ for a random structure of discs with reference to Figure 6.1. Following our treatment for random rectangles given in Section 4.2.3, the point autocorrelation function for coverage at points separated by a distance r is given by the ratio of the number of discs covering two points separated by a distance r to that of the number of discs covering one point. Referring to Figure 6.1, we observe that for any disc to cover both points p and q, its centre must lie within the shaded region. The area of the sector with angle θ is $R^2\,\theta/2$ and the area of the triangle AqB is $R^2\sin(\theta)/2$ so the shaded region has area

$$A_{\text{shaded}} = 2\left(\frac{R^2\,\theta}{2} - \frac{R^2}{2}\sin(\theta)\right) \tag{6.13}$$

$$= R^2\left(\theta - \sin(\theta)\right) \ . \tag{6.14}$$

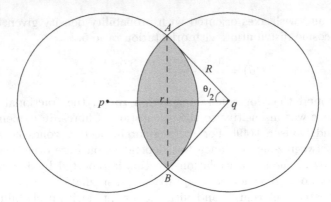

Figure 6.1. For a disc with diameter $D_d = 2R$ to cover both points p and q separated by a distance r, its centre must lie within the shaded area. (R.R. Farnood, C.T.J. Dodson and S.R. Loewen. *J. Pulp Pap. Sci.* **21**(10):J348-J356, 1995. Reproduced with permission)

From trigonometry, we obtain

$$\theta = 2 \arccos\left(\frac{r}{2R}\right) . \tag{6.15}$$

Substitution of Equation 6.15 into 6.13 yields, on simplification

```
In[1]:=  Ashaded = R² (θ - Sin[θ]);
         θ = 2 ArcCos[r / (2 R)];
         Ashaded = TrigExpand[Ashaded /. R → (Dd / 2)]
```

$$Out[3]= -\frac{1}{2} Dd\, r \sqrt{1 - \frac{r^2}{Dd^2}} + \frac{1}{2} Dd^2\, ArcCos\left[\frac{r}{Dd}\right]$$

The autocorrelation function for a random structure of discs is given by

$$\alpha(r) = \frac{A_{\text{shaded}}}{A_{\text{disc}}}, \tag{6.16}$$

so we have

```
In[4]:=  α[r_] := Simplify[Ashaded / (π Dd² / 4)]
         α[r]
```

$$Out[5]= \frac{-2r\sqrt{1 - \frac{r^2}{Dd^2}} + 2 Dd\, ArcCos\left[\frac{r}{Dd}\right]}{Dd\, \pi}$$

So the autocorrelation function is defined for $0 \le r \le D_d$ such that when $r = 0$, $\alpha(r) = 1$ and when $r \ge 1$, $\alpha(r) = 0$:

In[6]:= **Plot[α[r] /. Dd → 1, {r, 0, 1},**
AxesLabel → {"r/D$_d$", "α(r)"}]

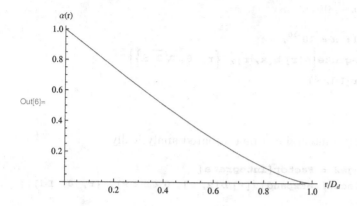

Note that by considering a structure of discs rather than rectangles, the geometry we consider is greatly simplified and we obtain a simple function for $\alpha(r)$ instead of the three-part piecewise function obtained for rectangles and given by Equation 4.44.

The fractional between-zones variance for a structure of random discs is given by

$$\rho_x = \int_\Omega \alpha(r)\, b(r)\, dr\,, \tag{6.17}$$

where the subscript x denotes that the variance depends on the scale of inspection and $b(r)$ is the probability density of separation of pairs of points r, as given by Equation 3.10 derived in Section 3.2.2:

In[7]:= **b[x_, r_] :=**
    Piecewise$\Big[\Big\{\Big\{$2 r $\left(\pi x^2 - 4\, x\, r + r^2\right)/x^4$, 0 < r ≤ x$\Big\}$,
        $\Big\{$2 r $\left(4\, x\, \sqrt{r^2 - x^2} - r^2 - x^2\, (2 + \pi - 4\, \text{ArcSin}[x/r])\right)\Big/x^4$,
        x < r ≤ $\sqrt{2}$ x$\Big\}\Big\}\Big]$

To determine the fractional between-zones variance, we must evaluate Equation 6.17 for two cases. For $0 \le x < D_d/\sqrt{2}$ we require numerical integration:

In[8]:= **Dd = 1; x = 0.5;**

NIntegrate$\left[\alpha[r]\ b[x,\ r],\ \left\{r,\ 0,\ \sqrt{2}\ x\right\}\right]$

Out[9]= 0.674618

and $\rho_x \to 1$ as $x \to 0$:

In[10]:= **Dd = 1; x = 10^{-10};**

NIntegrate$\left[\alpha[r]\ b[x,\ r],\ \left\{r,\ 0,\ \sqrt{2}\ x\right\}\right]$

Clear[Dd, x]

Out[11]= 1.

For $x \geq D_d/\sqrt{2}$ the integral can be evaluated analytically:

In[13]:= **ρxRange2 = Factor[Integrate[**
PiecewiseExpand[$\alpha[r]$ b[x, r], 0 < r < x], {r, 0, Dd}]]

Out[13]= $\dfrac{Dd^2\left(45\ Dd^2\ \pi - 1024\ Dd\ x + 180\ \pi^2\ x^2\right)}{720\ \pi\ x^4}$

In the original derivation of this fractional between-zones variance, Farnood *et al.* [50] approximated $b(r)$ and $\alpha(r)$ by polynomial functions in order to make the integrals tractable; their resultant expression for ρ_x is

$$\rho_x \approx \begin{cases} 1 - 0.592\ \frac{x}{D_d} & \text{for } 0 \leq x \leq 0.843\,D_d \\ \left(0.71 - 0.298\ \frac{D_d}{x}\right)\ \frac{D_d^2}{x^2} & \text{for } x > 0.843\,D_d \end{cases} \tag{6.18}$$

Plotting Equation 6.18 and the expression we have derived without approximation against x/D_d, we observe that the approximation of Farnood *et al.*, as represented by the broken line, slightly overestimates ρ_x for x/D_d less than about 1.

```
In[14]:= ρxapprx = Piecewise[{{1 - .592 x / Dd, 0 ≤ x ≤ 0.843 Dd},
            {(.71 - .298 Dd / x) (Dd / x)², x > 0.843 Dd}}];
        Dd = 1;
        Show[Plot[ρxRange2, {x, 1/√2, 5}],
          Plot[ρxapprx, {x, 0, 5},
            Exclusions → None, PlotStyle → Dashed],
          ListPlot[Prepend[
              Table[{x, NIntegrate[α[r] b[x, r], {r, 0, √2 x}]},
                {x, 0.05, .7, .05}], {0, 1}],
            PlotStyle → AbsolutePointSize[5]],
          PlotRange → All, AxesLabel → {"x/Dd", "ρx"}]
        Dd = .
```

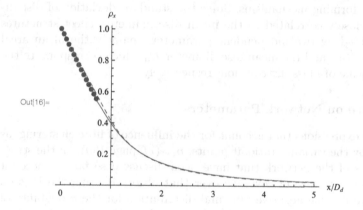

The variance of local areal density, $\sigma_x^2(\tilde{\beta})$ of a structure with mean areal density $\tilde{\beta}$ arising from the random deposition of discs is given by

$$\sigma_x^2(\tilde{\beta}) = \tilde{\beta}\,\beta_{\text{disc}}\,\rho_x ,\qquad(6.19)$$

where β_{disc} is the mass per unit area of a constituent disc.

The theory we have considered so far assumes all discs to have uniform diameter. Farnood *et al.* extended their theory to account for a distribution of disc diameters. Observing that the mass fraction of discs with diameter $D_{d,1} \leq D_d \leq D_{d,2}$ is proportional to

$$\int_{D_{d,1}}^{D_{d,1}} D_d^2\,f(D_d)\,\mathrm{d}D_d$$

where $f(D_d)$ is the probability density of disc diameters, the net fractional between-zones variance for a random structure of discs with a distribution of diameters $D_d^{\min} \leq D_d \leq D_d^{\max}$ is,

$$\rho_x^* = \frac{\int_{D_d^{\min}}^{D_d^{\max}} D_d^2 \, \rho_x \, f(D_d) \, \mathrm{d}D_d}{\int_{D_d^{\min}}^{D_d^{\max}} D_d^2 \, f(D_d) \, \mathrm{d}D_d} \tag{6.20}$$

Using a numerical stochastic decomposition technique, Farnood *et al.* [49, 51] compared the theory for the decay of variance with zone size for structures formed from discs with a lognormal distribution of diameters with those obtained experimentally by analysis of calibrated contact β-radiographs. The study considered 130 machine-made and laboratory formed networks, which were mainly paper samples but included also non-woven glass-fibre filters and historical data reported by Corte [18] from 22 papers manufactured in the 1970s. They reported a coefficient of determination of 0.99 for the correlation,

$$\sigma(D_d) = 1.2 \, \bar{D}_d^{\frac{2}{3}} \ . \tag{6.21}$$

The result is important since the data on which the regression was performed span 25 years and the correlation is applicable across a range of commercial and laboratory forming mechanisms. Since the standard deviation of disc diameters is so closely correlated to the mean disc diameter, sheet structures may be quantified by two independent parameters, namely the mean areal density of discs $\bar{\beta}_{\mathrm{disc}}$ and the mean disc diameter $\bar{D}_d$ which correspond to the intensity and scale of mass distribution, respectively.

6.3.1 Influence on Network Parameters

Here we derive expressions that account for the influence of fibre clustering, as characterised by the variance ratio at points, n_{vr} (*cf.* page 130), on the structural properties of the network that have been discussed so far. To account for the increased variance of coverage over that observed using the Poisson distribution, we use the negative binomial distribution for the probability of coverage. This distribution has been widely applied in other fields where data exhibit higher degrees of clustering, and hence higher variance, than those arising from equivalent Poisson processes, see *e.g.* [25, 46, 133].

The mean areal density of a network is invariably a product specification and is easily controlled during manufacturing. This means that for a given fibre type, the mean coverage also is implicitly specified and controlled. The variance of coverage is less easily controlled and we have established that in real networks it is typically greater than that calculated for a network of the same constituent fibres with their centres distributed according to a point Poisson process in two dimensions. The negative binomial distribution allows the variance of coverage at points to be varied independently of the mean, thus capturing the influence of process variables on network uniformity in a way that the Poisson distribution cannot. It has probability function,

In[1]:= **PDF[NegativeBinomialDistribution[n, p], c]**

Out[1]= $(1 - p)^c p^n$ Binomial $[-1 + c + n, -1 + n]$

The negative binomial distribution is often used to describe the distribution of the number of failures before n successes in a sequence of Bernoulli trials with success probability p. The binomial coefficient, and hence the probability function for the negative binomial distribution, can be expressed in terms of the Euler gamma function, **Gamma**:

In[2]:= **FunctionExpand [**
     PDF [NegativeBinomialDistribution [n, p], c]]

Out[2]= $\dfrac{(1 - p)^c p^n \text{ Gamma}[c + n]}{\text{Gamma}[1 + c] \text{ Gamma}[n]}$

The mean, variance and coefficient of variation are functions of the parameters n and p:

In[3]:= **Mnb = Mean[NegativeBinomialDistribution [n, p]]**
     Vnb = Variance[NegativeBinomialDistribution [n, p]]
     CVnb = PowerExpand $\left[\sqrt{\text{Vnb}} \,/\, \text{Mnb}\right]$

Out[3]= $\dfrac{n\,(1 - p)}{p}$

Out[4]= $\dfrac{n\,(1 - p)}{p^2}$

Out[5]= $\dfrac{1}{\sqrt{n}\,\sqrt{1 - p}}$

Parameters n and p do not lend themselves easily to interpretation in terms of the structure of the networks we seek to model. We therefore reparametrise the negative binomial probability function in terms of the mean coverage of the network $\bar{c}$ and the variance ratio $n_{vr} \geq 1$ such that the network has variance of coverage at points $\sigma^2(c) = n_{rv}\,\bar{c}$:

In[6]:= **np = Solve [{cbar == Mnb, (nvr cbar) == Vnb}, {n, p}] [[1]]**

Out[6]= $\left\{ n \to \dfrac{\text{cbar}}{-1 + \text{nvr}},\ p \to \dfrac{1}{\text{nvr}} \right\}$

```
In[7]:= P[nvr_, cbar_, c_] :=
        Evaluate[FullSimplify[FunctionExpand[
          PDF[NegativeBinomialDistribution[n, p], c] /. np],
          {c ∈ Integers, c ≥ 0 }]]
        P[nvr, cbar,
        c]
```

$$\text{Out[8]=} \quad \frac{\left(\frac{1}{nvr}\right)^{\frac{cbar}{-1+nvr}} \left(\frac{-1+nvr}{nvr}\right)^{c} \text{Gamma}\left[c + \frac{cbar}{-1+nvr}\right]}{c! \ \text{Gamma}\left[\frac{cbar}{-1+nvr}\right]}$$

Note that in the limit, as $n_{vr} \rightarrow 1$, we recover the probability function for the Poisson distribution:

```
In[9]:= Limit[P[nvr, cbar, c], nvr → 1,
        Direction → -1, Assumptions → {cbar, c} > 0]
```

$$\text{Out[9]=} \quad \frac{cbar^{c} \ e^{-cbar}}{c!}$$

The influence of our parameter n_{vr} on the distribution of local coverage is illustrated by comparing the probability density for the negative binomial distribution with $n_{vr} = 2$ for a network with $\bar{c} = 10$ (dashed line) with that for a random, i.e. Poissonian, network with the same mean coverage (solid line):

```
In[10]:= cbar = 10;
         randomcoverage = Table[
             {c, PDF[PoissonDistribution[cbar], c]}, {c, 0, 30}];
         clusteredcoverage = Table[
             {c, P[2, cbar, c]}, {c, 0, 30}];
         Show[ListPlot[{randomcoverage, clusteredcoverage},
           Filling → Axis, PlotStyle → AbsolutePointSize[4]],
         ListLinePlot[{randomcoverage, clusteredcoverage},
           PlotStyle → {{}, Dashed}], AxesLabel → {"c", "P(c)"}]
         cbar =.
```

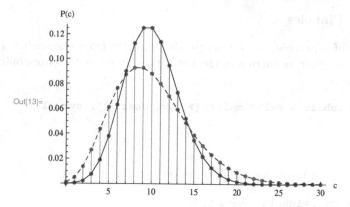

We observe also that whereas the skewness of the distribution decreases with increasing mean coverage, it increases as the variance ratio increases, so the distribution has a longer tail and the probability of points having low coverage increases.

```
In[15]:= sk = PowerExpand[Simplify[
        Skewness[NegativeBinomialDistribution[n, p] /. np]]]
      Plot3D[sk, {nvr, 1, 5}, {cbar, 1, 10},
      AxesLabel → {"n_vr", "c̄", "Sk"}]
```

$$\text{Out[15]=}\quad \frac{-1 + 2\,\text{nvr}}{\sqrt{\text{cbar}}\,\sqrt{\text{nvr}}}$$

Out[16]= Sk

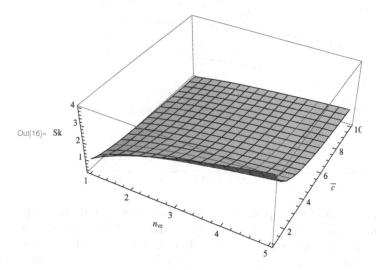

Probability of Pinholes

The probability of a pinhole, *i.e.* a void passing directly from one surface of the network to the other, occurring in the network is given by the probability of coverage zero:

In[17]:= **ProbPinhole = FullSimplify[P[nvr, cbar, 0], nvr > 1]**

Out[17]= $nvr^{-\frac{cbar}{-1+nvr}}$

so that we recover our result for random networks when $n_{vr} \to 1$:

In[18]:= **Limit[ProbPinhole, nvr → 1]**

Out[18]= e^{-cbar}

We plot the probability of pinholes for networks of mean coverage $\bar{c} = 5$ (dashed line), $\bar{c} = 10$ (solid line) and $\bar{c} = 15$ (dotted line). As expected, we observe that the probability of pinholes decreases as mean coverage increases and increases as n_{vr} increases:

In[19]:= **LogPlot[{ProbPinhole /. cbar → 5,**
   ProbPinhole /. cbar → 10, ProbPinhole /. cbar → 15},
   {nvr, 1, 4}, PlotStyle → {Dashed, {}, Dotted},
   AxesLabel → {"n$_{vr}$", "P(0)"}]

Out[19]=

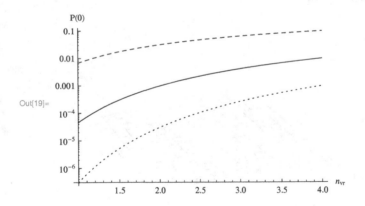

Although the probability of pinholes is very low, their occurrence can still be significant in, for example, filtration applications. If we assume that the diameter of a circular pinhole is of order that of a fibre, say 10 μm, we can readily show that if the probability of pinholes $P(0)$ is of order 10^{-6}, then the expected number of pinholes per square metre is of order 10^4:

In[20]:= **PinholeArea** = $\pi \left(10 \ 10^{-6}\right)^2 \Big/ 4.$ (*m²*);

ProbPinholes = 10^{-6};

nPinholes = **ProbPinholes** / **PinholeArea**

Out[22]= 12 732.4

Inter-crossing Distances

To derive the distribution of inter-crossing distances, we assume that the negative binomial distribution describes the frequency of crossings along a line with μ crossings per unit length. The expected number of crossings in an interval of length g is μg and the probability of there being no crossings in a gap of length g is

In[23]:= **Pzero** = **Simplify[P[nvr, μ g, 0], nvr > 1]**

Out[23]= $\mathrm{nvr}^{-\frac{g\,\mu}{-1+\mathrm{nvr}}}$

The probability density of g is given by

In[24]:= **pdfg** = **FullSimplify[Pzero / Integrate[Pzero,**
        {g, 0, ∞}, Assumptions → nvr > 1 && μ > 0]]

Out[24]= $\dfrac{\mathrm{nvr}^{-\frac{g\,\mu}{-1+\mathrm{nvr}}}\ \mu\ \mathrm{Log[nvr]}}{-1+\mathrm{nvr}}$

with mean

In[25]:= **gb** = **Integrate[g pdfg,**
        {g, 0, ∞}, Assumptions → nvr > 1 && μ > 0]

Out[25]= $\dfrac{-1+\mathrm{nvr}}{\mu\ \mathrm{Log[nvr]}}$

Solving for μ and substituting into our probability density **pdfg** we obtain the probability density for inter-crossing distances in terms of the mean inter-crossing distance only:

In[26]:= **Solve[gb == gbar, μ][[1]]**
        pdfg = **FullSimplify[pdfg /. %]**

Out[26]= $\left\{\mu \to \dfrac{-1 + \text{nvr}}{\text{gbar Log[nvr]}}\right\}$

Out[27]= $\dfrac{e^{-\frac{g}{\text{gbar}}}}{\text{gbar}}$

> The resultant expression is, of course, the probability density function for the exponential distribution. We conclude then that the distribution of inter-crossing distances is not influenced by clustering.

As crossings can occur on both sides of a given fibre, intervals containing a small integer number of crossings may be more appropriate for characterisation of the inter-fibre crossing distances that form the in-plane perimeter of pores within the sheet [29]. The distribution of intervals containing n crossings in a fibre network is

$$f(g,n) = \frac{P(n)}{\int_0^\infty P(n)\,dg} \quad . \tag{6.22}$$

We consider first the *random* case where $P(n)$ is given by the Poisson distribution. The probability density $f(g,n)$ is given by

In[28]:= **pdfgn = PDF[PoissonDistribution[μ g], n] /**
  Integrate[PDF[PoissonDistribution[μ g], n],
  {g, 0, ∞}, Assumptions $\to$ n > 0 && μ > 0]

Out[28]= $\dfrac{e^{-g\,\mu}\,\mu\,(g\,\mu)^n}{n!}$

with mean

In[29]:= **Integrate[g pdfgn, {g, 0, ∞}, Assumptions $\to$ n > 0 && μ > 0]**

Out[29]= $\dfrac{1+n}{\mu}$

Assuming that expected length of an interval containing n crossings is $\overline{g_n} = (n+1)\,\bar{g}$ we have $\bar{g} = 1/\mu$. We obtain the final form of the probability density function for the distances of intervals containing n crossings by making this substitution:

In[30]:= **pdfgn =**
       PowerExpand[FunctionExpand[pdfgn /. $\mu \to (1 / gbar)$]]

Out[30]= $\dfrac{e^{-\frac{g}{gbar}} \, g^n \, gbar^{-1-n}}{Gamma[1 + n]}$

We note that this is the probability density function for the gamma distribution with mean, $\overline{g_n} = (n+1)\,\bar{g}$ and coefficient of variation, $1/\sqrt{n+1}$:

In[31]:= **PDF[GammaDistribution[n + 1, gbar], g]**

Out[31]= $\dfrac{e^{-\frac{g}{gbar}} \, g^n \, gbar^{-1-n}}{Gamma[1 + n]}$

The corresponding closed-form expression for the probability density of intervals containing n crossings in a *clustered* fibre network cannot be obtained analytically. The probability densities for some small n can be obtained in closed form however, and for $n = 1$ we recover the gamma distribution which we have just shown to describe the intervals for the random case:

In[32]:= **P1 = FullSimplify[P[nvr, μ g, 1], nvr > 1];**
       pdfg1 = P1 / Integrate[P1,
           {g, 0, ∞}, Assumptions $\to$ nvr > 1 && μ > 0];
       gb = Integrate[g pdfg1, {g, 0, ∞},
           Assumptions $\to$ nvr > 1 && μ > 0];
       Solve[gb == (2 gbar), μ][[1]];
       pdfg1 = FullSimplify[pdfg1 /. %]
       TrueQ[pdfg1 == pdfgn /. n $\to$ 1]

Out[36]= $\dfrac{e^{-\frac{g}{gbar}} \, g}{gbar^2}$

Out[37]= True

For $n = 2$ and $n = 3$ the probability densities become increasingly unwieldy:

In[38]:= **P2 = FullSimplify[P[nvr, μ g, 2], nvr > 1];**
       pdfg2 = P2 / Integrate[P2,
           {g, 0, ∞}, Assumptions $\to$ nvr > 1 && μ > 0];
       gb = Integrate[g pdfg2, {g, 0, ∞},
           Assumptions $\to$ nvr > 1 && μ > 0];
       Solve[gb == (3 gbar), μ][[1]];
       pdfg2 = FullSimplify[pdfg2 /. %]

$$
\text{Out[42]=} \left(4\, e^{-\frac{2\,g\,(3+\text{Log[nvr]})}{3\,\text{gbar}\,(2+\text{Log[nvr]})}}\, g\, (3 + \text{Log[nvr]})^2 \right.
$$

$$
\left. (6\,g + \text{Log[nvr]}\ (2\,g + 6\,\text{gbar} + 3\,\text{gbar}\,\text{Log[nvr]})) \right) \Big/
$$

$$
\left(27\,\text{gbar}^3\,(2 + \text{Log[nvr]})^4 \right)
$$

```
In[43]:=  P3 = FullSimplify[P[nvr, μ g, 3], nvr > 1];
          pdfg3 = P3 / Integrate[P3,
             {g, 0, ∞}, Assumptions → nvr > 1 && μ > 0];
          gb = Integrate[g pdfg3, {g, 0, ∞},
             Assumptions → nvr > 1 && μ > 0];
          Solve[gb == (4 gbar), μ][[1]];
          pdfg3 = FullSimplify[pdfg3 /. %]
```

$$
\text{Out[47]=} \left(e^{-\frac{g\,(12+\text{Log[nvr]}\,(9+2\,\text{Log[nvr]}))}{4\,\text{gbar}\,(3+\text{Log[nvr]})\,(3+\text{Log[nvr]}))}}\, g\, (12 + \text{Log[nvr]}\ (9 + 2\,\text{Log[nvr]}))^2 \right.
$$

$$
(12\,g + \text{Log[nvr]}\ (9\,g + 12\,\text{gbar} +
$$
$$
2\,\text{Log[nvr]}\ (g + 6\,\text{gbar} + 2\,\text{gbar}\,\text{Log[nvr]})))
$$
$$
(12\,g + \text{Log[nvr]}\ (9\,g + 24\,\text{gbar} + 2\,\text{Log[nvr]}
$$
$$
\left. (g + 12\,\text{gbar} + 4\,\text{gbar}\,\text{Log[nvr]}))) \right) \Big/
$$
$$
\left(512\,\text{gbar}^4\,(3 + \text{Log[nvr]}\ (3 + \text{Log[nvr]}))^5 \right)
$$

Given our earlier analysis, the coefficient of variation of intervals containing n crossings is known analytically for the cases when $n = 0$ and $n = 1$; for the cases when $n = 2$ and $n = 3$ we must compute the coefficient of variation numerically. Inevitably, the coefficient of variation decreases with increasing n, since by including more crossings in the intervals of interest, we effectively apply an averaging filter to the process of crossings on a line. Interestingly, the dependence on n_{vr} is rather weak after an initial increase:

```
In[48]:= gbar = 1;
         cvg0 = 1;
         cvg1 = 1 / √2 ;
         cvg2 = Table[
             {nvr, √(NIntegrate[ (g - 3 gbar)² pdfg2, {g, 0, ∞}]) /
                 (3 gbar)}, {nvr, 1, 5, .1}];
         cvg3 = Table[{nvr,
                 √(NIntegrate[ (g - 4 gbar)² pdfg3, {g, 0, ∞}]) /
                 (4 gbar)}, {nvr, 1, 5, .1}];
         Show[Plot[{cvg0, cvg1}, {nvr, 1, 5},
             PlotStyle → {{}, Dashed}],
           ListLinePlot[{cvg2, cvg3},
             PlotStyle → {Dotted, DotDashed}],
           PlotRange → All, AxesLabel → {"n_vr", "cv(g_n)"}]
         gbar =.
```

We observe also that the probability density of intervals containing n crossings (solid lines) closely resembles that of a gamma distribution with the same mean and variance (dashed lines). Here we illustrate this for the intervals containing $n = 2$ and $n = 3$ crossings when $n_{vr} = 3$:

In[55]:= **nvr = 3; gbar = 1;**

$$\mathbf{cv2} = \sqrt{\mathbf{NIntegrate}\left[(g - 3\,gbar)^2\,pdfg2,\ \{g,\ 0,\ \infty\}\right]} \,\Big/$$

    (3 gbar);

$$\mathbf{cv3} = \sqrt{\mathbf{NIntegrate}\left[(g - 4\,gbar)^2\,pdfg3,\ \{g,\ 0,\ \infty\}\right]} \,\Big/$$

    (4 gbar);

Plot[{pdfg2,
    PDF[GammaDistribution[1/cv2², 3 gbar cv2²], g], pdfg3,
    PDF[GammaDistribution[1/cv3², 4 gbar cv3²], g]},
    {g, 0, 10}, PlotStyle → {{}, Dashed},
    AxesLabel → {"gₙ", "pdf(gₙ)"}]

Out[58]=

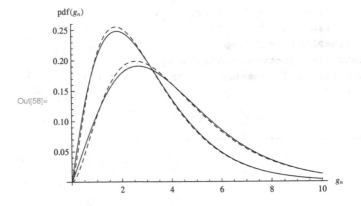

Recall that the inter-crossing distances represent the boundaries of voids in the plane of the network. Despite an expectation that pore size will be influenced by network uniformity [20, 27, 38], experimental evidence suggests that any dependence is rather weak [35]. Our earlier treatments suggest two reasons for this, firstly that the adjacent sides of polygons representing voids in the plane are correlated, and that this correlation is expected to be rather insensitive to fibre orientation and clustering (*cf.* Section 3.3.4), and secondly that measurements of pore size are strongly influenced by the out-of-plane dimensions of voids (*cf.* Section 5.3), these being rather insensitive to network uniformity [36, 66]. The insensitivity of the distribution of inter-crossing distances and the distributions of intervals containing n crossings to the parameter n_{vr}, which quantifies clustering, provides a third contributing factor to the observed insensitivity of pore size to network uniformity. Interestingly, we observe that the dependence of the coefficient of variation of inter-crossing distances is most sensitive to n_{vr} when it is close to 1; any influence in near-random structures would therefore be rather difficult to isolate experimentally.

Absolute Contact States

We derived expressions for the absolute contact states for random networks, following Kallmes *et al.* [76] in Section 4.3.4. The absolute contact states represent the fractions of the fibre surface in the network that are available for contact with no other fibres, one other fibre or two other fibres. For clustered networks, we repeat their derivation using the negative binomial distribution for the probability of coverage *c*. Here we consider only the first stage of the treatment provided in Section 4.3.4, and assume that the probability of contact between vertically adjacent fibres is 1 such that we consider the fraction of the fibre surface *available* for contact with zero, one or two other fibres.

We begin by defining the probability density function for the negative binomial distribution in terms of the variance ratio, n_{vr}.

In[1]:= **P[c_] :=** $\left((-1 + nvr)^c \right.$
   nvr$^{-(c + cbar/(nvr-1))}$ **Gamma[c + cbar / (nvr - 1)]** $\left. \right) /$
   (c ! Gamma[cbar / (-1 + nvr)])

The fraction of the total fibrous length which does not make contacts with other fibres is that with coverage 1, divided by the mean coverage:

In[2]:= **C0 = FullSimplify[P[1] / cbar]**

Out[2]= $nvr^{-1-\frac{cbar}{-1+nvr}}$

At points with coverage greater than 1, the two outermost fibres are in contact with other fibres on one side only and this fraction is

In[3]:= **C1 = FullSimplify[2 (1 - P[0] - P[1]) / cbar]**

Out[3]= $\dfrac{2 \left(1 - nvr^{-1-\frac{cbar}{-1+nvr}} (cbar + nvr) \right)}{cbar}$

Similarly, at points with coverage greater than 2, all except the two outermost fibres make contact on both sides:

In[4]:= **C2 = FullSimplify[**
   Sum[(c - 2) P[c] / cbar, {c, 3, ∞}], cbar > 0 && nvr > 1]

Out[4]= $\dfrac{-2 + cbar + nvr^{-1-\frac{cbar}{-1+nvr}} (cbar + 2 \, nvr)}{cbar}$

such that $C_0 + C_1 + C_2 = 1$:

In[5]:= **TrueQ[Simplify[C0 + C1 + C2] == 1]**

Out[5]= True

and we recover the expressions for the random case in the limit as $n_{vr} \to 1$:

In[6]:= **Limit[{C0, C1, C2}, nvr → 1]**

Out[6]= $\left\{ e^{-cbar}, \dfrac{2\, e^{-cbar}\left(-1 - cbar + e^{cbar}\right)}{cbar}, \right.$

$\left. \dfrac{e^{-cbar}\left(2 + cbar - 2\, e^{cbar} + cbar\, e^{cbar}\right)}{cbar} \right\}$

Plots of the absolute contact states against the variance ratio for networks with differen mean coverage $\bar{c}$ show a strong dependence of the absolute contact states on the extent of clustering when the mean coverage is low, and a diminishing dependence as coverage increases:

In[7]:= **Plot3D[C0, {nvr, 1, 4}, {cbar, 1, 10},**
      AxesLabel → {"n_{vr}", "c̄", "C(0)"}, PlotRange → All]
    Plot3D[C1, {nvr, 1, 4}, {cbar, 1, 10},
      AxesLabel → {"n_{vr}", "c̄", "C(1)"}]
    Plot3D[C2, {nvr, 1, 4}, {cbar, 1, 10},
      AxesLabel → {"n_{vr}", "c̄", "C(2)"}]

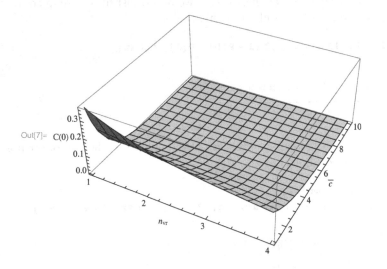

Out[7]=

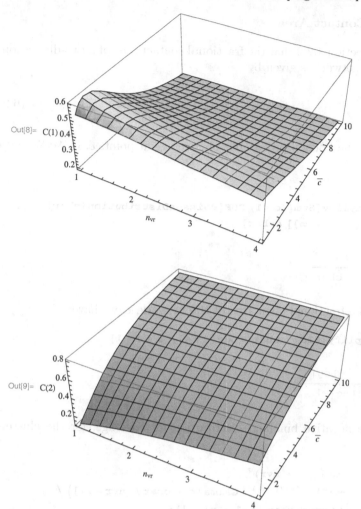

The absolute contact states are most sensitive to clustering at low mean coverages. Here we observe that increasing n_{vr} decreases $C(0)$ and $C(1)$ as a greater fraction of fibre length exists in the bulk of the sheet; accordingly the fraction $C(2)$ increases. At higher coverages, the fraction $C(2)$ is approximately constant though we see a small decrease in $C(1)$ which is associated with a similar increase in $C(0)$. The influence of clustering becomes negligible when $\bar{c}$ is greater than about 10.

When $\bar{c}$ is greater than about 10, the fraction $C(0) \approx 0$; the fractions $C(1)$ and $C(2)$ may be considered independent of n_{vr} and are approximated by $2/\bar{c}$ and $(1 - 2/\bar{c})$, respectively.

Fractional Contact Area

Recall from Section 4.2.2 that the fractional contact area of a two-dimensional random fibre network is given by

$$\Phi_{2D} = \frac{1}{\bar{c}} \sum_{c=1}^{\infty} (c - 1)\, P(c), \tag{6.23}$$

where $P(c)$ is the Poisson probability of coverage at points c. In *Mathematica* we obtain

```
In[1]:= Φ2d =
        Simplify[Sum[(c - 1) PDF[PoissonDistribution[cbar], c],
            {c, 1, ∞}] / cbar]
```

$$Out[1]= 1 + \frac{-1 + e^{-cbar}}{cbar}$$

In terms of the fractional open area of the network, ϵ, we have:

```
In[2]:= Simplify[Φ2d /. cbar → Log[1 / ε], 0 < ε < 1]
```

$$Out[2]= \frac{1 - \epsilon + Log[\epsilon]}{Log[\epsilon]}$$

Using the negative binomial distribution for coverage, for the clustered case we have

```
In[3]:= P[c_] := ((-1 + nvr)^c
            nvr^-(c + cbar/(nvr-1)) Gamma[c + cbar / (nvr - 1)]) /
            (c! Gamma[cbar / (-1 + nvr)]);
        Φ2d = Apart[FullSimplify[
            Sum[(c - 1) P[c], {c, 1, ∞}] / cbar, nvr > 1 && cbar > 0]]
```

$$Out[4]= \frac{-1 + cbar}{cbar} + \frac{nvr^{-\frac{cbar}{-1 \cdot nvr}}}{cbar}$$

We recover the result for the random case as $n_{vr} \to 1$:

```
In[5]:= Limit[Φ2d, nvr → 1]
```

$$Out[5]= 1 + \frac{-1 + e^{-cbar}}{cbar}$$

The fractional open area of our two-dimensional clustered network is given by the probability of coverage zero:

In[6]:= **P[0]**

Out[6]= $nvr^{-\frac{cbar}{-1-nvr}}$

We therefore have the following unified expression for the fractional contact area of random and clustered two-dimensional networks:

$$\Phi^*_{2D} = 1 - \frac{1}{\bar{c}} + \frac{\epsilon}{\bar{c}} \, . \qquad (6.24)$$

For two-dimensional networks of equal coverage, the free parameter ϵ in Equation 6.24 increases with the degree of clustering, so the fractional contact area also increases. Such behaviour is consistent with expectation, since clustering will increase the fraction of the network with coverage greater than 2 and hence the proportion of the total fibre length that contacts other fibres on both sides. Given that the fractional contact area of our networks can be stated in terms of the fractional open area and the mean coverage only, it follows that the fractional contact area of a clustered network with mean coverage $\bar{c}$ is the same as that of a random network with the same fractional open area and mean coverage given by:

In[7]:= **Simplify[cbarclustered /.**
  Solve[(P[0] /. cbar → cbarclustered) == e^{-cbar},
  cbarclustered][[1]], cbar > 0]

Out[7]= $\dfrac{cbar\ (-1 + nvr)}{Log[nvr]}$

This is very convenient since it means that the fractional contact area of a multi-planar structure of infinite coverage is the same for clustered networks as for a random network; accordingly, we may use the expression for Φ_∞ derived on page 134:

In[8]:= **Φinf = 1 + (ϵ (2 + (ϵ − 3) ϵ) / Log[ϵ]);**

When we consider multi-planar networks of finite thickness we must take account of the fraction of the total fibre length which is located at the surfaces of the network and that can be in contact with other fibres on one side only. At points with coverage c, the fraction of fibre surfaces available for contact with other fibres is $(c-1)/c$. The fraction of the network covered by fibres is $(1 - P(0))$ so the fraction of the network as a whole that is available for contact is,

In[9]:= **f = FullSimplify[(1 / (1 - P[0]))**
　　　　Sum[((c - 1) / c) P[c], {c, 1, ∞}], cbar > 0 && nvr > 1]

Out[9]= $1 - \dfrac{\text{cbar HypergeometricPFQ}\left[\left\{1, 1, 1 + \frac{\text{cbar}}{-1+\text{nvr}}\right\}, \{2, 2\}, \frac{-1+\text{nvr}}{\text{nvr}}\right]}{\text{nvr}\left(-1 + \text{nvr}^{\frac{\text{cbar}}{-1+\text{nvr}}}\right)}$

> The fractional contact area of a clustered network with finite coverage $\bar{c}$ is less than that of a random network of infinite coverage by a factor, f. This is dependent on the mean coverage of the network and the variance ratio n_{vr}.

We observe that this fraction is most sensitive to clustering at low mean coverages where a greater fraction of the total coverage at points is in the surfaces of the network:

In[10]:= **Plot3D[f, {nvr, 1, 5}, {cbar, 0, 20},**
　　　　AxesLabel → {"c̄", "n_vr", "f" }]

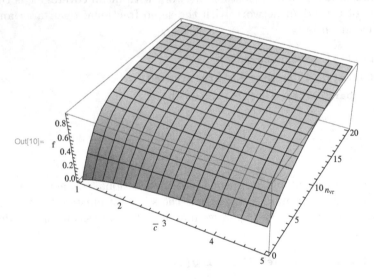

Out[10]=

A plot of the surface representing the fractional contact area Φ_c against variance ratio and the normalised density $(1 - \epsilon)$ shows that the fractional contact area is most sensitive to clustering at higher densities, where fibre contact is greatest:

```
In[11]:=  Φc = f Φinf;
          cbar = 5;
          ParametricPlot3D[{nvr, (1 - ε), Φc},
           {nvr, 1, 5}, {ε, 0, 1}, BoxRatios → {1, 1, .4},
           AxesLabel → {"nvr", "(1-ε)", "Φc"}]
          cbar = .
```

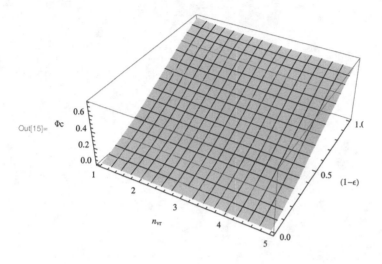

Out[15]=

7

Three-dimensional Networks

7.1 Introduction

So far we have considered two classes of idealised planar fibre networks: those where fibre centres occur in two dimensions only and multi-planar networks consisting of superimposed two-dimensional structures, so that they are representative of layered structures. In Chapters 3 to 6 we have developed these models to characterise the structures of those stochastic fibrous materials where the principal axes of fibres lie within a few degrees of the plane of the network. A less common, but nonetheless important class of stochastic fibrous materials are those where fibre axes may be oriented at angles that are manifestly not in the plane of the material, should this be definable. Materials of this type include needled and hydro-entangled non-woven textiles, acoustic and thermally insulating waddings, and metallic fibre networks for application in prosthetics and metal-matrix composites.

Tomographic images of sintered three-dimensional metal fibre networks are shown in Figure 7.1. The networks shown were formed by compression of loosely packed stainless steel fibres of length 5 mm and diameter 60 μm within a tube of diameter 15 mm [158]. On first inspection it is clear that in the lightly compressed network shown on the left of Figure 7.1, fibres are oriented at a range of angles to the xy-plane; on further compression fibre axes lie closer to the xy-plane, as illustrated in the image on the right of Figure 7.1. If the network were to be compressed to its limit, fibres would lie with their axes close to the plane of the network, though the structure of the material would differ from the multi-planar networks that we have considered earlier in that fibres may extend from one surface to another. Thus, full compression of a three-dimensional structure into a near-planar structure would yield a fully felted, rather than a layered, material.

Three-dimensional fibre networks are important also in the manufacture of paper and wet- and air-laid non-woven webs, since these are formed from the filtration of three-dimensional fibrous suspensions. Although these suspensions are a precursor to the materials that interest us here, they are widely studied

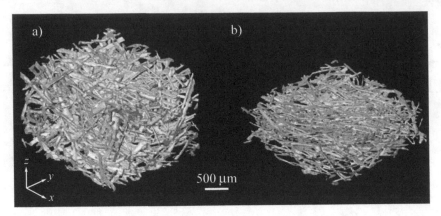

Figure 7.1. Tomographic images of three-dimensional sintered networks of stainless steel fibres with mean width, $\omega = 60$ μm and length $\lambda = 5$ mm. a) Approximately isotropic network with mean porosity $\epsilon \approx 0.9$; b) compressed network with mean porosity $\epsilon \approx 0.5$. (J.C. Tan, J.A. Elliott and T.W. Clyne. Analysis of tomography images of bonded fibre networks to measure distributions of fibre segment length and fibre orientation. *Adv. Eng. Mater.* **8**(6):495-500, 2006. Copyright Wiley-VCH Verlag GmbH & Co. KGaA. Reproduced with permission)

in the context of these manufacturing processes since they exhibit interesting rheologies and their structure is dynamic, with stochastic flow fields resulting in dispersion and flocculation of the structure. Accordingly, network structure and rheology are interdependent and contribute strongly to the extent that the structure of the suspension persists in the resultant near-planar network. For a recent review of suspension rheology with emphasis on network structure see [84].

The classical reference structure for materials of this type is a random three-dimensional fibre network where fibre centres are distributed according to a point Poisson process in three dimensions and fibre axes have a uniform distribution within the solid angle 2π.

We begin our treatment using *Mathematica* to generate a graphical representation of a point Poisson process of 1,000 fibre centres within a unit cube:

```
In[1]:= n = 1000;
       SeedRandom[1]
       centres = RandomReal[{0, 1}, {n, 3}];
       ListPointPlot3D[centres, BoxRatios → {1, 1, 1}]
```

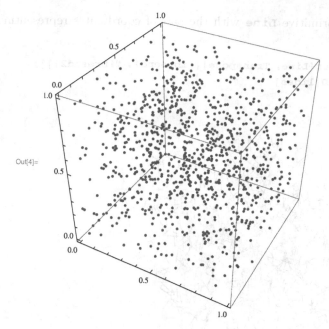

Out[4]=

In a two-dimensional rendering such as that generated, it is difficult to appreciate the inherent clustering of fibre centres that is characteristic of our Poisson process. However, the output we have generated can be rotated in *Mathematica* using a mouse or other input device, aiding our perception of the relative locations of the points in three-dimensions.

To extend our process from points to lines of finite length, we generate the random angles θ and ϕ that represent the orientation of the lines to the x-y and x-z planes. These angles are uniformly distributed between $-\pi$ and π.

```
In[5]:= θ = RandomReal[{-π, π}, n];
       φ = RandomReal[{-π, π}, n];
```

To generate a graphic of random lines with finite length we specify the line length λ and compute the coordinates of their ends. For our example, we generate lines of length 0.1:

```
In[7]:= λ = .1;
       fibreend1 = centres - Transpose[
           λ {Cos[θ] Cos[φ], Sin[θ] Cos[φ], Sin[φ]} / 2];
       fibreend2 = centres + Transpose[
           λ {Cos[θ] Cos[φ], Sin[θ] Cos[φ], Sin[φ]} / 2];
```

The graphic we require is obtained using **Graphics3D** where this command operates on a list of lines obtained by applying the command **Map** to associate the graphics primitive **Line** with the pair of coordinates representing the fibre ends:

In[10]:= **lines = Map[Line, Transpose[{fibreend1, fibreend2}]];
Graphics3D[lines]**

Out[11]=

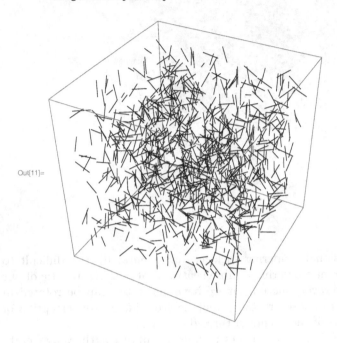

7.2 Network Density

When considering planar and layered fibre networks, we used the dimensionless variables coverage and porosity to characterise our structures. For planar random networks of fibres we saw that the two-dimensional analogue of porosity, the fractional open area, ϵ was dependent on the mean coverage, $\bar{c}$ through the Poisson probability of coverage zero:

$$\epsilon = e^{-\bar{c}} .$$

For layered fibre networks, porosity can be varied independently of coverage.

In the three-dimensional case, the property analogous to coverage is the volumetric concentration of fibres, C_v as defined by the porosity of the network:

$$C_v = (1 - \epsilon) , \tag{7.1}$$

such that C_v is a dimensionless density. Accordingly, the volumetric concentration of a network with density ρ formed from fibres with density ρ_f is

$$C_v = \frac{\rho}{\rho_f} . \tag{7.2}$$

When considering suspensions of fibres, the volume of the network occupied by solid is often difficult to determine experimentally and it is therefore more convenient to consider the mass per unit volume of the suspension, which we denote C_m (kg m^{-3}):

$$C_m = C_v \rho_f . \tag{7.3}$$

Of course, for dry fibres $C_m = \rho$, but in wet-laying processes, fibres in suspension are often swollen by the process water that constitutes the suspending medium. Also, many natural fibres are hollow, so the volume occupied by fibres including intra-fibre liquid increases. If the dry mass of fibre per unit volume is C_m and fraction of the swollen fibre mass that consists of liquid is k, then the volumetric concentration is given by

$$C_v = C_m \left(\frac{1}{\rho_f} + \frac{k}{\rho_l} \right) , \tag{7.4}$$

where ρ_l is the density of the suspending liquid.

Most three-dimensional fibre networks have volumetric concentrations of less than about 10 % and for flowing suspensions in fibre handling and web-forming processes, concentrations are more commonly around 1 to 2 %. To reach very high concentrations, fibres must be sufficiently flexible to conform to each other and this raises the interesting problem as to what fibre parameters influence the maximum packing density of rigid fibres. This problem was addressed by Parkhouse and Kelly [120] who considered the probability that an additional rod could be randomly placed within an existing network without being obstructed by other rods. For rigid rods of aspect ratio, A, Parkhouse and Kelly derived the maximum packing concentration as

$$C_v^{\max} = \frac{2 \log(A)}{A} . \tag{7.5}$$

Their analysis utilised empirical relationships obtained from simulation studies, so we do not repeat the derivation here. Parkhouse and Kelly observed good agreement between Equation 7.5 and experimental data for rods of aspect ratio greater than about 6. An alternative derivation was provided by Evans and Gibson [48] who found:

$$C_v^{\max} = \frac{k}{A} . \tag{7.6}$$

From their theoretical treatment, Evans and Gibson obtained $k = 4$, though better agreement with experimental data was observed when $k = 5.3$. By considering the statistics of fibre contact, Philipse [122] also obtained a relationship of the form given by Equation 7.6 and from experimental data obtained $k = 5.4 \pm 0.2$. Philipse provides the physical interpretation of k as the expected number of contacts per fibre in a three-dimensional network and we will consider this property in Section 7.3.

The important property obtained by Evans and Gibson [48], Parkhouse and Kelly [120] and Philipse [122] is that the maximum packing concentration for random three-dimensional networks of rigid rods depends only on their aspect ratio. If three-dimensional networks are to be formed at higher concentrations, then we require that fibres are either flexible or the criteria for randomness are broken. An example of the latter case is provided by Evans and Gibson, who note that in the dynamics of liquid crystals, exceeding the maximum packing concentration provides favourable conditions for transition to nematic structures.

7.2.1 Crowding Number

A convenient and useful form of the volumetric concentration of three-dimensional fibre networks is the expected number of fibres in a sphere of diameter one mean fibre length. This parameter was proposed by Kerekes and co-workers [81–83] and termed the *crowding number*. The networks of interest to Kerekes *et al.* were flowing fibrous suspensions, which flocculate to give a greater degree of clustering than that observed for a point Poisson process of fibre centres in three-dimensions. We derive the crowding number, n_{crowd} by considering the volumetric concentration of the network.

The volume of a single fibre of length λ and width ω is

$$V_{fib} = \frac{\pi \omega^2 \lambda}{4} \,, \tag{7.7}$$

and the volume of a sphere of diameter λ is

$$V_{sphere} = \frac{\pi \lambda^3}{6} \,. \tag{7.8}$$

Denoting the expected number of fibres in a sphere of diameter one mean fibre length, n_{crowd}, the volumetric concentration, C_v is given by

$$C_v = \frac{n_{crowd} V_{fib}}{V_{sphere}} \,. \tag{7.9}$$

Solving for n_{crowd} in *Mathematica* we obtain,

```
In[1]:=  Vfib = π ω² λ / 4;
         Vsphere = π λ³ / 6;
         Solve[Cv == (ncrowd Vfib / Vsphere), ncrowd][[1, 1]]


Out[3]=  ncrowd → 2 Cv λ²
                  ─────────
                   3 ω²
```

In terms of the fibre aspect ratio, $A = \lambda/\omega$, we have:

In[4]:= `% /. λ → (ω A)`

Out[4]= $\text{ncrowd} \to \dfrac{2\,A^2\,Cv}{3}$

We have noted that the volumetric concentration is often difficult to determine and it is convenient to express the crowding number in terms of the mass concentration, C_m. For a network of fibres of length λ and coarseness δ, the mass of a single fibre is $\lambda\,\delta$, so we have:

In[5]:= `Mfib = λ δ;`
`Solve[Cm == (ncrowd Mfib / Vsphere), ncrowd][[1, 1]]`

Out[6]= $\text{ncrowd} \to \dfrac{Cm\,\pi\,\lambda^2}{6\,\delta}$

Kerekes *et al.* [81] identify three ranges of n_{crowd} that they associate with the propensity of a fibre suspension to flocculate. Inevitably, if $n_{\text{crowd}} \leq 1$, then fibres are expected to be independent of each other, so we can expect fibre centres to be distributed according to a point Poisson process in three dimensions. Arguably, the term 'network' should not be applied when $n_{\text{crowd}} \leq 1$, but at higher crowding numbers fibres interact with each other and networks do form. Kerekes and Schell [82] suggest that fibres are in continuous mutual contact when the crowding number exceeds 60; more recently, Sampson [140] suggested this transition occurs when $n_{\text{crowd}} = A$ and, developing the treatment of Doi and Edwards [43, 44], Kerekes proposes that the threshold crowding number is $n_{\text{crowd}} = A/2$. The absolute value of this threshold is less important than its consequence for the structure of fibrous suspensions. This being that above some critical crowding number, fibre interactions result in the suspension exhibiting a greater degree of clumping than would be found in the corresponding Poisson process in three dimensions. An alternative threshold crowding number was identified by Martinez *et al.* [98] using positron emission tomography. They identified a 'gel'-concentration, below which the presence of fibres has limited influence on suspension rheology, and determined that this concentration occurs when $n_{\text{crowd}} \approx 16$. Note also that correlations have been reported between suspension crowding in papermaking systems and the variance ratio of the resultant sheets [28, 42, 86, 99].

The concentration when $n_{\text{crowd}} = 1$ provides a measure analogous to the percolation threshold for a three-dimensional network. Rigorously, we require a percolation threshold to provide a criterion for connectedness, but when fibres are free to move, as they are in a suspension, the concentration when $n_{\text{crowd}} = 1$ is useful as it provides the criterion for independence of fibres. The problem was first addressed by Mason [100] using the same considerations as those on which we based our derivation of the crowding number. This critical concentration is given in terms of the volumetric and mass concentration by,

In[7]:= **Cvcrit = Cv /. Solve$\left[(2\, A^2\, Cv\, / \, 3) == 1,\ Cv \right] [[1]]$**

CmCrit = Cm /. Solve$\left[(Cm\, \pi\, \lambda^2 \, / \, (6\, \delta)) == 1,\ Cm \right] [[1]]$

Out[7]= $\dfrac{3}{2\, A^2}$

Out[8]= $\dfrac{6\, \delta}{\pi\, \lambda^2}$

respectively.

7.3 Intensity of Contacts

To obtain the expected number of contacts per fibre in a three-dimensional network we allow rods to penetrate each other such that a contact is considered to occur when any part of the volume occupied by one fibre passes through that occupied by another fibre. This assumption greatly simplifies our analysis and has been used widely in the literature [33, 122, 161]. Inevitably the assumption introduces some error, though Toll [161] points out that if fibres are displaced such that they just make contact, then intersections will be generated with the same probability that they are broken.

The derivation presented here draws upon that provided by Dodson [33], which in turn was based on the methods of van Wyk [164]. We seek an expression for the expected number of fibres that intersect the volume of a single cylindrical fibre of length, λ and diameter ω.

Consider a cubic volume of side λ that contains a fibre network of volumetric concentration C_v. The volume of a fibre is $V_{\text{fib}} = \pi\, \omega^2\, \lambda/4$ and the expected number of fibres occurring within a cubic volume of side λ is

$$n_v = \frac{C_v\, \lambda^3}{V_{\text{fib}}} \,. \tag{7.10}$$

Consider now a projection of the fibre network onto the base of the cube. The expected number of fibres *per unit area* in the projected network, n_{fib} is:

In[1]:= **Vfib = $\pi\, \omega^2\, \lambda$ / 4;**

nfib = Cv λ / Vfib

Out[2]= $\dfrac{4\,Cv}{\pi\,\omega^2}$

A fibre oriented at angle θ to the vertical projects a length $\lambda_p = \lambda \sin(\theta)$ onto the plane. In Section 4.2.1 we saw that the probability density of the angles between pairs of lines was $\sin(\theta)$, so the expected projected length of fibres with all orientations, $\overline{\lambda_p}$ is:

In[3]:= **λpbar = Integrate[λ Sin[θ] Sin[θ], {θ, 0, π / 2}]**

Out[3]= $\dfrac{\pi\,\lambda}{4}$

and the total projected fibre length per unit area is

In[4]:= **λppua = nfib λpbar**

Out[4]= $\dfrac{Cv\,\lambda}{\omega^2}$

Refer now to Figure 7.2 which shows a reference fibre within the cubic volume that we are considering. For another fibre to make contact with this fibre, its centreline must pass through the cylindrical volume of diameter $2\,\omega$ that includes the reference fibre. The projected area of this cylindrical volume is $\pi\,\omega^2$, so the expected amount of projected fibre length occurring within this projected area is

In[5]:= **λpcircle = λppua π ω²**

Out[5]= $Cv\,\pi\,\lambda$

The centreline of each fibre passing through the projected circle generates a chord within that circle and the expected length of random chords drawn in a circle of radius ω is $\pi\omega/2$. The expected number of fibres with part of their volume passing through the reference fibre provides our expression for the expected number of contacts per fibre:

In[6]:= **Ncontacts = λpcircle / (π ω / 2)**

Out[6]= $\dfrac{2\,Cv\,\lambda}{\omega}$

In[7]:= **Ncontacts /. λ → (A ω)**

Out[7]= $2\,A\,Cv$

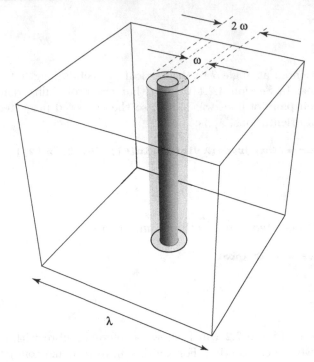

Figure 7.2. Reference fibre of length λ and diameter ω within a cubic volume of side one fibre length. Fibres with centrelines passing through the cylindrical volume of diameter 2ω will intersect the reference fibre

Our final expression therefore tells us that the expected number of contacts per fibre is proportional to the volumetric concentration of the network and the aspect ratio of the fibres. This seems intuitively reasonable and we note that the same expression was obtained independently by Toll [161], Philipse [122] and Dodson [33]. Modified theories of fibre contact have been derived which seek to account for the reduction in the volume available for additional contacts due to that occupied by existing contacts, see *e.g.* [87, 116, 117]; to a reasonable degree of approximation however, the proportionalities to aspect ratio and volumetric concentration persist over the typical range of application.

7.4 Variance of Porosity

In Section 5.6.2 we used the distribution of porosity in planar fibrous materials to calculate the distribution of local average flow rates using the Kozeny-Carman equation. These same calculations hold for three-dimensional networks, but the dependence of the distribution of porosity on fibre variables for three-dimensional networks differs from that of planar and multi-layered networks. Here we derive the variance of porosity following the treatment of Dodson and Sampson [39].

Consider a three-dimensional random fibre network with volumetric concentration C_v formed from fibres with aspect ratio A. In the previous section, we derived the expected number of crossings per fibre in such a network as

$$\bar{n}_c = 2 A C_v . \tag{7.11}$$

By definition, the mean porosity of the network $\bar{\epsilon} = 1 - \bar{C}_v$. So, denoting the local average porosity $\tilde{\epsilon}$, the local average number of contacts per fibre is

$$\tilde{n}_c = 2 A (1 - \tilde{\epsilon}) . \tag{7.12}$$

When discussing local average properties and their variability, we should specify the scale of inspection. An appropriate inspection volume for our three-dimensional networks is a cube of side x, since this allows complete sampling of the volume by contiguous cubic volumes. From Equation 7.12 it follows that the variance of the local average number of crossings per fibre is

$$\sigma_x^2(\tilde{n}_c) = \sigma_x^2 (2 A (1 - \tilde{\epsilon}))$$
$$= 4 A^2 \sigma_x^2(\tilde{\epsilon}) , \tag{7.13}$$

where the subscript x denotes the scale of inspection. The variance of local average porosity is therefore given by

$$\sigma_x^2(\tilde{\epsilon}) = \frac{\sigma_x^2(\tilde{n}_c)}{4 A^2} . \tag{7.14}$$

Now, Equation 7.11 tells us the expected number of crossings on a single fibre in a random three-dimensional network. If the number of crossings per fibre is a Poisson distributed variable, then the variance of the number of crossings on individual fibres is

$$\sigma_0^2(\tilde{n}_c) = \bar{n}_c . \tag{7.15}$$

The local average number of fibres in cubic volumes of side x will vary from region to region, but if the average number is $\bar{N}_f$, then from the central limit theorem, the variance of averages of samples of size $\bar{N}_f$ is $1/\bar{N}_f$ times the variance of samples of size 1, $i.e.$

$$\sigma_x^2(\tilde{n}_c) = \frac{1}{\bar{N}_f} \sigma_0^2(\tilde{n}_c)$$
$$= \frac{\bar{n}_c}{\bar{N}_f} \tag{7.16}$$

The expected number of fibres in a volume x^3 is

```
In[1]:=  Vfib = π ω² λ / 4;
         Nf = Cv x³ / Vfib
```

$$Out[2]= \frac{4\ Cv\ x^3}{\pi\ \lambda\ \omega^2}$$

and from Equation 7.16, the variance of the local average number of contacts per fibre is

```
In[3]:=  A = λ / ω;
         nc = 2 A Cv;
         Varxnc = nc / Nf
```

$$Out[5]= \frac{\pi\ \lambda^2\ \omega}{2\ x^3}$$

We may now obtain the variance of local porosity using Equation 7.14:

```
In[6]:=  varxe = Varxnc / (4 A²)
```

$$Out[6]= \frac{\pi\ \omega^3}{8\ x^3}$$

So, at a given scale of inspection, the variance of local porosity is proportional to the cube of the fibre diameter only. Importantly, it is independent of the mean porosity, and thus it is independent of the volumetric concentration also.

Of course, if the variance of local porosity is independent of the mean porosity, then the same is not true for the coefficient of variation of local porosity. When generating a plot of this parameter as a function of inspection zone size x and the volumetric concentration, we bear in mind that our derivation requires $\bar{N}_f > 1$. The following example is plotted for fibres of aspect ratio $A = 100$ and all other lengths are given as multiples of a fibre width.

```
In[7]:=  A = 100;
         λ = A ω;
         ω = 1;
         Plot3D[If[Nf > 1, √varxe / (1 - Cv)],
           {x, 0, 20}, {Cv, 0, .6}, PlotPoints → 75,
           PlotRange → All, AxesLabel → {"x", "Cᵥ", "CVₓ(ε)"}]
         Clear[A, λ, ω]
```

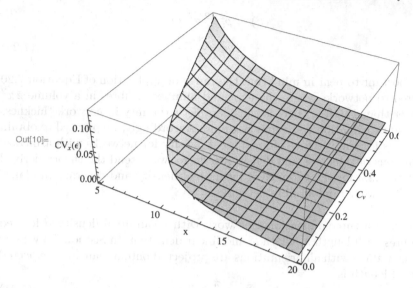

As expected, we observe that the coefficient of variation of local porosity decreases with increasing inspection zone size and increases with increasing volumetric concentration.

7.5 Variance of Areal Density

A projection of the three-dimensional fibre onto its plane of support yields the distribution of mass of the network in two-dimensions. For a network of mean porosity $\bar{\epsilon}$ and mean thickness $\bar{z}$ formed from fibres with density ρ_f, the mean areal density is given by

$$\bar{\beta} = \bar{z}\,(1 - \bar{\epsilon})\,\rho_f \ . \tag{7.17}$$

We have derived the variance of local average porosity for contiguous cubic volumes of side x as

$$\sigma_x^2(\bar{\epsilon}) = \frac{8}{\pi}\,\frac{\omega^3}{x^3} \ . \tag{7.18}$$

If we partition the plane of support of the network into square zones of side x, then the local average thickness of the network above these square zones is $\tilde{z}$. If the variance of local average thickness is small, then $\tilde{z} \approx \bar{z}$ and the between-zones variance of porosity, as obtained by partitioning our network into contiguous cuboids with base x^2 and height, $\bar{z}$ is

$$\sigma_x^2(\tilde{\epsilon}) = \frac{8}{\pi}\,\frac{\omega^3}{\bar{z}\,x^2} \ . \tag{7.19}$$

From Equation 7.17, the variance of local areal density is approximated by

$$\sigma_x^2(\tilde{\beta}) \approx \rho_f^2 \, \bar{z}^2 \, \sigma_x^2(\tilde{\epsilon})$$

$$\approx \rho_f^2 \, \bar{z} \, \frac{8}{\pi} \frac{\omega^3}{x^2} \; . \tag{7.20}$$

It is important to bear in mind that the range of application of Equation 7.20 is limited to networks where the expected number of fibres in a volume $\bar{z} x^2$ is greater than 1. Often there is considerable variability in network thickness and the approximation $\tilde{z} \approx \bar{z}$ may not hold. Accordingly, we proceed to obtain expressions for the variance of local areal density for networks with arbitrary distributions of thickness and density; to do so, we extend the theory derived in Section 4.2.3 for the variance of local areal density and coverage in planar and multi-planar random fibre networks.

Consider a three-dimensional network with mean areal density $\bar{\beta}$ formed from fibres with length λ, width ω and linear density δ. In Section 7.3 we saw that when fibres with all orientations are projected onto a plane, the expected projected length is

$$\bar{\lambda}_p = \frac{\pi}{4} \lambda \; . \tag{7.21}$$

It follows that the expected mass per unit projected length of the fibres is

$$\bar{\delta}_p = \frac{4}{\pi} \delta \; , \tag{7.22}$$

and the expected mass per unit projected area of the fibres is

$$\bar{\beta}_{f,p} = \frac{\bar{\delta}_p}{\omega}$$

$$= \frac{4\,\delta}{\pi\omega} \; . \tag{7.23}$$

The mean coverage of the network is

$$\bar{c} = \frac{\bar{\beta}}{\bar{\beta}_{f,p}}$$

$$= \frac{\pi}{4} \frac{\omega}{\delta} \, \bar{\beta} \; . \tag{7.24}$$

Recall that the random variable coverage is distributed according to a point Poisson process in two dimensions, so the variance of coverage at points is given by:

$$\sigma_p^2(c) = \bar{c}$$

$$= \frac{\pi}{4} \frac{\omega}{\delta} \, \bar{\beta} \tag{7.25}$$

In Section 4.2.3 we obtained the variance of coverage at points of a random planar or multi-planar fibre network as

$$\sigma^2(c) = \frac{\omega}{\delta}\,\bar{\beta}\;. \tag{4.27}$$

So, the variance of coverage at points of a random three-dimensional network is less than that of a random planar or multi-planar fibre network with the same mean areal density formed from the same constituent fibres by a factor $\pi/4$.

The variance of areal density at points is given by:

$$
\begin{aligned}
\sigma_p^2(\beta) &= \sigma_p^2(c\,\bar{\beta}_{f,p})\\
&= \bar{\beta}_{f,p}^2\,\sigma_p^2(c)\\
&= \frac{4\,\delta}{\pi\omega}\,\bar{\beta}
\end{aligned}
\tag{7.26}
$$

Again, in Section 4.2.3 we obtained the variance of areal density at points of a random planar or multi-planar fibre network as

$$\sigma^2(\beta) = \frac{\delta}{\omega}\,\bar{\beta}\;. \tag{4.28}$$

So, the variance of areal density at points of a random three-dimensional network is greater than that of its corresponding random planar or multi-planar fibre network by a factor $4/\pi$.

The between-zones variance of coverage and areal density are obtained by multiplying the point variances by the fractional between-zones variance for a projected network $\rho_{x,p}$, i.e.

$$\sigma_{x,p}^2(\tilde{c}) = \sigma_p^2(c)\,\rho_{x,p} \tag{7.27}$$
$$\sigma_{x,p}^2(\tilde{\beta}) = \sigma_p^2(\beta)\,\rho_{x,p}\;. \tag{7.28}$$

We derived expressions for the fractional between-zones variance for random planar and near-planar networks in Section 4.2.3 and showed that it is a function of fibre length, λ, fibre width, ω and inspection zone size, x:

$$\rho_x = \int_0^{\sqrt{2}\,x} \alpha(r,\lambda,\omega)\,b(r,x)\;\mathrm{d}r\;, \tag{7.29}$$

where $\alpha(r,\lambda,\omega)$ is the point autocorrelation function and $b(r,x)$ is the probability density function for the separation of pairs of points. In a projection of a three-dimensional network onto a plane, fibre centres are distributed according to a point Poisson process in two dimensions, so the probability density function for the separation of pairs of points is $b(r,x)$ unaltered. The projected width of fibres onto the plane is unaltered also, so the point autocorrelation function is influenced only by the reduction in effective fibre length due to its projection onto the plane, and we have

$$\rho_{x,p} = \int_0^{\sqrt{2}\,x} \alpha(r, \lambda_p, \omega)\, b(r, x)\, dr \ . \tag{7.30}$$

We obtained expressions for $b(r, x)$ and $\alpha(r, \lambda, \omega)$ in Sections 3.2.2 and 4.2.3 respectively, so input these to *Mathematica* in the form presented there:

In[1]:=
```
b[x_, r_] :=
    Piecewise[{{2 r (π x² - 4 x r + r²) / x⁴, 0 < r ≤ x},
    {2 r (4 x √(r² - x²) - r² - x² (2 + π - 4 ArcSin[x / r])) / x⁴,
    x < r ≤ √2 x}}]
```

In[2]:=
```
α[λ_, ω_, r_] :=
    Piecewise[{{1 - (2 r)/(π λ) - (2 r)/(π ω) + r²/(π λ ω), 0 ≤ r ≤ ω},
    {-(2 r)/(π ω) - ω/(π λ) + (2 √(r² - ω²))/(π ω) + (2 ArcSin[ω/r])/π, ω < r ≤ λ},
    {(2 √(r² - λ²))/(π λ) - r²/(π λ ω) - λ/(π ω) - ω/(π λ) + (2 √(r² - ω²))/(π ω) -
    (2 ArcCos[λ/r])/π + (2 ArcSin[ω/r])/π, λ < r ≤ √(λ² + ω²)}}]
```

We define a function to calculate ρ_x as given by Equation 7.29:

In[3]:=
```
ρx[λ_, ω_, x_] :=
    NIntegrate[α[λ, ω, r] b[x, r], {r, 0, √2 x}]
```

From Equation 7.21, $\bar{\lambda}_p = \pi \lambda/4$, so the function to evaluate $\rho_{x,p}$ as given by Equation 7.30 differs from that for ρ_x only in the value of λ used in the evaluation of $\alpha(r, \lambda, \omega)$:

In[4]:=
```
ρxp[λ_, ω_, x_] :=
    NIntegrate[α[π λ / 4, ω, r] b[x, r], {r, 0, √2 x}]
```

The following code generates a plot of ρ_x (solid line) and $\rho_{x,p}$ (dashed line) against inspection zone size for fibres of length $\lambda = 2$ mm and width $\omega = 20$ μm.

```
In[5]:= xtab = Join[Range[.1, 2, .1], Range[2.5, 10, 0.5]];
        ρxtab =
          Table[ρx[2, .02, xtab[[i]]], {i, 1, Length[xtab]}];
        ρxptab = Table[ρxp[2, .02, xtab[[i]]],
          {i, 1, Length[xtab]}];
        ListLogPlot[{Transpose[{xtab, ρxtab}],
          Transpose[{xtab, ρxptab}]}, Joined → True,
          AxesLabel → {"x", "ρx, ρx,p"}, PlotStyle → {{}, Dashed}]
```

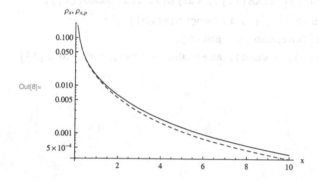

Out[8]=

We observe that as a consequence of the reduction in the contribution of fibre length to variance arising from its projection onto the plane, the fractional between-zones variance, $\rho_{x,p}$ is less than that for fibres lying in the plane of the network, ρ_x. Although the effect increases with increasing zone size, the fractional between-zones variance decays steeply with inspection zone size, such that the absolute difference is greatest at scales around a fibre length. Note that our treatment contains an approximation since we consider only the mean projected length $\bar{\lambda}_p$. However, we expect the influence of the distribution of λ_p on ρ_{x_p} to be of order 10 % or less [42].

To obtain the variance of local areal density in our three-dimensional network, $\sigma_{x,p}(\tilde{\beta})$, we multiply the fractional between-zones variance, $\rho_{x,p}$ by the variance of local areal density at points $\sigma_p(\beta)$ as given by Equation 7.26. We define also a function to compute the variance of local areal density $\sigma_x(\tilde{\beta})$ for planar and near-planar structures:

```
In[9]:= Varβ[λ_, ω_, x_, βbar_, δ_] :=
          δ βbar ρx[λ, ω, x] / (ω / 1000)
        Varβp[λ_, ω_, x_, βbar_, δ_] :=
          (4 / π) δ βbar ρxp[λ, ω, x] / (ω / 1000)
```

Mean areal density is typically reported with units $g\,m^{-2}$, so the factor 1,000 is included in the above code such that we obtain the variance of local areal density in units of $g^2\,m^{-4}$ when length, width and inspection zone size are specified in mm and coarseness is specified with units $g\,m^{-1}$. We compute and plot the variance of local areal density for three-dimensional (dashed line) and near-planar (solid line) networks with mean areal density $\bar{\beta} = 100 \text{ g m}^{-2}$ formed from fibres with length 2 mm, width 20 μm and coarseness $1.5 \times 10^{-4} \text{ g m}^{-1}$ as follows:

```
In[11]:= varβtab = Table[
            {xtab[[i]], Varβ[2, .02, xtab[[i]], 100, 1.5 10⁻⁴]},
            {i, 1, Length[xtab]}];
         varβptab = Table[{xtab[[i]], Varβp[2, .02, xtab[[i]],
            100, 1.5 10⁻⁴]}, {i, 1, Length[xtab]}];
         ListLinePlot[{varβtab, varβptab},
           PlotStyle → {{}, Dashed}, AxesLabel → {"x", "σ²ₓ, σ²ₓ,ₚ"}]
```

We observe that at scales of inspection less than about two fibre lengths, the variance of local areal density of our three-dimensional network is greater than that of a two-dimensional network formed from the same fibres at the same mean areal density. Given the sensitivity of the fractional between-zones variance to inspection zone size, the greater variance observed for three-dimensional structures is attributable to the greater variance at points in these networks.

We state then, that we expect the variance of areal density of three-dimensional structures to be $4/\pi$ times that of the corresponding planar or near-planar structure at points, and to be approximately the same at large scales of inspection.

7.6 Sphere Caging

Whereas for planar and multi-layered fibrous materials we defined the characteristic dimensions of inter-fibre voids with reference to the plane of the material, the characteristic dimensions of inter-fibre voids in three-dimensional

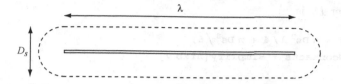

Figure 7.3. Planar representation of the excluded volume of a sphere of diameter D_s and a fibre of length λ. When the centre of a sphere lies within the volume represented by the broken line, contact occurs between the fibre and the rod

networks are more difficult to conceptualise. The fibre ligaments between contacts partition the volume into convex polyhedra; the polygonal faces of these polyhedra may be considered to represent constrictions through which particles, fluids, *etc.* can pass and we might expect these polygons to have the same distributions of sides, perimeter, area, *etc.* as arise from the planar line processes that we discussed in Chapter 3. An elegant approach to characterising the inter-fibre space in three-dimensional networks is provided by Philipse and Kluijtmans [123] who considered the caging of spheres so that they are immobilised through their contacts with fibres. Here we follow their treatment to derive the expected diameter of spheres caged by the network in this way.

Consider a network of rods of high aspect ratio such that $\lambda \gg \omega$ and a test sphere of diameter D. As in our derivation of the expected number of contacts per fibre in Section 7.3, we allow rods to penetrate the sphere such that contacts occur when any part of the volume occupied by one fibre passes through that occupied by the sphere. Assuming the width of the rod to be negligible, the volume within which the centre of the sphere must lie for it to make contact with a given rod, as illustrated in Figure 7.3, is given by

$$V = \frac{\pi}{4} D_s^2 \lambda + \frac{\pi}{6} D_s^3 . \tag{7.31}$$

The volume of a fibre is $V_{\text{fib}} = \pi \omega^2 \lambda / 4$, so in a network of volumetric concentration C_v, the expect number of fibres per unit volume is

$$n_{\text{fib}} = \frac{C_v}{V_{\text{fib}}} . \tag{7.32}$$

In *Mathematica* we obtain:

```
In[1]:= Vfib = π ω² λ / 4;
        nfib = Cv / Vfib

             4 Cv
Out[2]=   ─────────
           π λ ω²
```

We assume that fibre contacts are generated when the centre of our test sphere lies within the volume V, so the expected number of fibre contacts on a sphere of diameter D_s is

In[3]:= **V = π Ds^2 λ / 4 + π Ds^3 / 6;**
    Ncontacts = Simplify[nfib V]

Out[4]= $\dfrac{Cv\ Ds^2\ (2\ Ds + 3\ \lambda)}{3\ \lambda\ \omega^2}$

In systems of practical interest, we expect $\lambda \gg D_s$ and we obtain the approximation:

In[5]:= **Ncontacts = Limit[Ncontacts /. λ $\rightarrow$ (λonDs Ds), λonDs $\rightarrow$ ∞]**

Out[5]= $\dfrac{Cv\ Ds^2}{\omega^2}$

At least four contacts with fibres are required to cage a sphere. Whether this is a sufficient number of contacts to fully immobilise the sphere depends on the location of contacts relative to each other. So, for example, if a sphere is in contact with four fibres and these contacts all lie on one hemisphere, then the sphere is not immobilised. Typically then, we require more than four contacts to cage a sphere. To immobilise a sphere from movement in two dimensions, at least three contacts are required and Philipse and Kluijtmans [123] show analytically the expected number of contacts required is five. The expected number of contacts required to cage a sphere in three dimensions is not known analytically, so Philipse and Kluijtmans carried out simulations by increasing the number of contacts randomly located on the surface of a test sphere until at least one combination of four of these contacts caged the sphere according to a force balance. They found that up to 15 contacts were required to meet this criterion, and that the average number of contacts required to cage the sphere was 7. Using this number, we can obtain the expected diameter $\bar{D}_s$ of a sphere caged in our network:

In[6]:= **Ds = Ds /. Solve[Ncontacts == 7, Ds][[2]]**

Out[6]= $\dfrac{\sqrt{7}\ \omega}{\sqrt{Cv}}$

So the expected diameter of spheres caged within a network increases with fibre width and decreases with volumetric concentration. We note however that for a fixed number of fibres per unit volume, changing the width of fibres will change the volumetric concentration also. We obtain a more appropriate parametrisation by considering the total fibre length per unit volume,

$$\tau = n_{\text{fib}}\,\lambda \qquad\qquad (7.33)$$

We obtain the expected sphere diameter in terms of τ only:

In[7]:= **Solve[τ == nfib λ, Cv][[1]]**
PowerExpand[Ds /. %]

Out[7]= $\left\{ \text{Cv} \rightarrow \dfrac{1}{4}\,\pi\,\tau\,\omega^2 \right\}$

Out[8]= $\dfrac{2\sqrt{\dfrac{7}{\pi}}}{\sqrt{\tau}}$

In[9]:= **N[%]**

Out[9]= $\dfrac{2.98541}{\sqrt{\tau}}$

We may state then that the expected diameter of a sphere caged by fibres in a three-dimensional network is

$$\bar{D}_s \approx \frac{3}{\sqrt{\tau}}\,, \qquad\qquad (7.34)$$

and is therefore independent of fibre width and length and decreases as the total fibre length per unit volume increases.

Recall that we observed a similar dependency when considering the dimensions of polygons generated by random lines in the plane.

For completeness, we note the result of Ogston [113] who considered the distribution of the diameters of spheres contacting one fibre only and obtained,

$$\bar{D}_s = \sqrt{\frac{\pi}{4\,C_v}}\,\omega\,. \qquad\qquad (7.35)$$

Applying Ogston's criterion to the treatment of Philipse and Kluijtmans [123] we obtain:

In[10]:= **Clear[Ds]**
DsOgston = Ds /. Solve[Ncontacts == 1, Ds][[2]]

Out[11]= $\dfrac{\omega}{\sqrt{\text{Cv}}}$

This estimate differs from that given by Equation 7.35 by a factor $\sqrt{\pi/4} \approx 0.88$. The difference arises because Ogston did not allow fibres to penetrate spheres; though by permitting fibres and spheres to occupy the same volume, we greatly simplify the analysis. Ogston's treatment does provide however the distribution of the diameters of spheres contacting one fibre as,

$$f(D_s) = \frac{\pi D_s}{2 \bar{D}_s^2} e^{-\frac{\pi D_s^2}{4 \bar{D}_s^2}} \tag{7.36}$$

We obtain the variance and coefficient of variation of sphere diameters in the usual way:

In[12]:= **pdfDs** = $\left(\pi\, \text{Ds}\, /\, \left(2\, \text{Dbar}^2\right)\right)\, \text{e}^{-\pi\, \text{Ds}^2/\left(4\, \text{Dbar}^2\right)}$;

VarDs = **PowerExpand** $\left[\text{Integrate}\left[\left(\text{Ds} - \text{Dbar}\right)^2\, \text{pdfDs},\right.\right.$

$\left.\left.\{\text{Ds}, 0, \infty\}, \text{Assumptions} \rightarrow \text{Re}\left[\text{Dbar}^2\right] > 0\right]\right]$

CVDs = **PowerExpand** $\left[\sqrt{\text{VarDs}}\, /\, \text{Dbar}\right]$

Out[13]= $\dfrac{\text{Dbar}^2\, (4 - \pi)}{\pi}$

Out[14]= $\sqrt{\dfrac{4 - \pi}{\pi}}$

Note that the coefficient of variation obtained is independent of the mean suggesting that the probability density of sphere diameters should be well approximated by a gamma distribution. We plot the cumulative distribution functions with a solid line for Ogston's probability density and a dashed line for the corresponding gamma distribution:

In[15]:= **cdfDs** = **Integrate[pdfDs, {Ds, 0, Ds}]**;

Plot $\left[\left\{\text{cdfDs}\, /.\, \text{Dbar} \rightarrow 1,\right.\right.$

$\text{CDF}\left[\text{GammaDistribution}\left[1/\text{CVDs}^2, \text{CVDs}^2\right], \text{Ds}\right]\right\},$

$\{\text{Ds}, 0, 3\}, \text{PlotStyle} \rightarrow \{\{\}, \text{Dashed}\},$

$\left.\text{AxesLabel} \rightarrow \left\{"\text{Ds}/\bar{\text{D}}_\text{s}", "\text{F}(\text{D}_\text{s})"\right\}\right]$

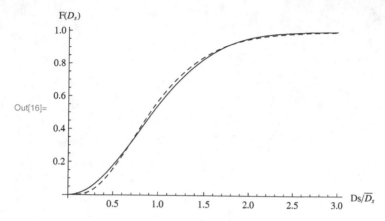

So, just as we saw for the when considering planar fibre processes, the characteristic dimensions of voids in three-dimensional networks are well characterised as being distributed according to a gamma distribution and the network variable that controls the mean is the expected total fibre length per unit volume.

Here and in earlier sections, we have considered fibres as rigid cylindrical rods. When considering flexible fibres these may be considered as consisting of straight segments with bends between them. The volume surrounding such fibres that we have considered here and in Section 7.3 is unaltered by the number of segments that constitute a fibre and their orientation. Accordingly, we expect no influence of fibre flexibility on the statistics we have computed; this remark was made also by Ogston [113]. It is supported by the fact that the total fibre length per unit volume, and not the length of individual fibres, is the controlling parameter defining void dimensions. So, in the context of sphere caging, the random orientation of short fibre segments should be equivalent to the random orientation of long straight rods with the same total fibre length per unit volume. However, it should be noted that in real structures formed from flexible fibres, consolidation of the structure will result in an increased incidence of fibre axes lying at angles close to the plane of the material, *i.e.* perpendicular to the direction of consolidation. To account for such effects new models must be derived.

References

1. M.S. Abdel-Ghani and G.A. Davies. Simulation of non-woven fibre mats and the application to coalescers. *Chem. Eng. Sci.* **40**(1):117–129, 1985
2. R. Amiri, J.R. Wood, A. Karnis, J. Görres. The apparent density of paper. *J. Pulp Pap. Sci.* **20**(5):142–148, 1994
3. C. Antoine, P. Nygard, Ø.W. Gregersen, R. Holmstad, T. Weitkamp and C. Rau. Three dimensional images of paper obtained by phase contrast X-ray microtomography: image quality and binarisation. *Nucl. Instr. Meth. Phys. Res. A.* **490**(1-2):392–402, 2002
4. M. Avikainen and A.L. Erkilla. Comparison of traditional beta-radiography and storage phosphor screen formation measurement techniques. *Pap. ja Puu* *85*(5):279–286, 2003
5. A. Baddeley. Stochastic geometry: An introduction and reading list. *Int. Statist. Rev.* **50**(2):179–193, 1982
6. I. Balberg, and N. Binenbaum. Computer study of the percolation threshold in a two-dimensional anisotropic system of conducting sticks. *Phys. Rev. B.* **28**(7):3799–3812, 1983
7. I. Balberg, N. Binenbaum and C.H. Anderson. Critical behavior of the two-dimensional sticks system. *Phys. Rev. Lett.* **51**(8):1605–1608, 1983
8. C. Bettstetter, H. Hartenstein and X. Pérez-Costa. Stochastic properties of the random waypoint mobility model. *Wireless Networks* **10**(5):555 567, 2004
9. W.C. Bliesner. A study of the porous structure of fibrous sheets using permeability techniques. *Tappi J.* **47**(7):392–400, 1964
10. P.A. Boeckerman. Meeting the special requirements for on-line basis weight measurement of lightweight nonwoven fabrics. *Tappi J.* **75**(12):166–172, 1992
11. R.C. Brown. The pore size distribution of model filters produces by random fragmentation described in terms of the Weibull distribution. *Chem. Eng. Sci.* **49**(1):145–146, 1994
12. J.P. Casey. **Pulp and Paper Chemistry and Chemical Technology**, Vol. 2. Third edition. John Wiley and Sons, New York, 1980
13. J. Castro and M. Ostoja-Starzewski. Particle sieving in a random fiber network. *Appl. Math. Modelling* **24**(8-9):523–534, 2000
14. C. Chatfield. **Statistics for Technology**. Third edition. Chapman and Hall, London, 1983

15. X. Cheng and A.M. Sastry. On transport in stochastic, heterogeneous fibrous domains. *Mech. Mater.* **31**(12):765–786, 1999

16. L.A. Clarenburg and H.W. Piekaar. Aerosol filters–I. Theory of the pressure drop across single component glass fibre filters. *Chem. Eng. Sci.* **23**(7):765–771, 1968

17. L.A. Clarenburg and H.W. Piekaar. Aerosol filters–II. Theory of the pressure drop across multi-component glass fibre filters. *Chem. Eng. Sci.* **23**(7):773–781, 1968

18. H. Corte. Über die Verteilung der Massendichte in Papier – Zweiter teil: Ergebnisse an füllstoffreinen Papieren. (On the distribution of mass density in paper – part II: Results on unfilled papers). *Das Papier* **24**(5):261–264, 1970

19. H. Corte and O.J. Kallmes. Statistical geometry of a fibrous network. In **Formation and Structure of Paper** *Trans. IInd Fund. Res. Symp.* (F. Bolam, ed.), pp. 13–52, BPBMA, 1962

20. H. Corte and E.H. Lloyd. Fluid flow through paper and sheet structure. In **Consolidation of the Paper Web** *Trans. IIIrd Fund. Res. Symp.* (F. Bolam, ed.), pp. 981–1009, BPBMA, London, 1966

21. H.L. Cox. The elasticity and strength of paper and other fibrous materials. *Brit. J. Appl. Phys.* **3**:72–9, 1952

22. I.K. Crain and R.E. Miles. Monte Carlo estimates of the distributions of the random polygons determined by random lines in the plane. *J. Statist. Comput. Simul.* **4**:293–325, 1976

23. T. Cresson. The sensing, analysis and simulation of paper formation. PhD Thesis, State University of New York, 1982

24. T.M. Cresson, H. Tomimasu, P. Luner. Characterisation of paper formation. Part 1: sensing paper formation. *Tappi J.* **73**(7):153–159, 1990

25. D.J. Croton, M. Colless, E. Gaztanaga *et al.* The 2dF galaxy redshift survey: voids and hierarchical scaling models. *Month. Not. Roy. Astron. Soc.* **352**(3):828–836, 2004

26. M.F. Dacey. Some properties of order distance for random point distributions. *Geografiska Annaler B* **49**(1):25–32, 1967

27. M. Deng and C.T.J. Dodson. **Paper: An Engineered Stochastic Structure**. Tappi Press, Atlanta, 1994

28. M. Deng and C.T.J. Dodson. Random star patterns and paper formation. *Tappi J.* **77**(3):195–199, 1994

29. R.W. Dent. Inter-fiber distances in paper and nonwovens. *J. Text. Inst.* **92**(1):63–74, 2001

30. C.T.J. Dodson. A contribution to the statistical rheology of bonded fibrous networks. PhD thesis, Brunel University, 1969

31. C.T.J. Dodson. Spatial variability and the theory of sampling in random fibrous networks. *J. Roy. Statist. Soc.* **B 33**(1):88–94, 1971

32. C.T.J. Dodson. On the distribution of pore heights in random layered fibre networks. In **The Science of Papermaking** (C.F. Baker, ed.), *Trans. XIIth Fund. Res. Symp.*, pp. 1037–1042, FRC, Manchester, 2001

33. C.T.J. Dodson. Fiber crowding, fiber contacts and fiber flocculation. *Tappi J.* **79**(9):211–216, 1996

34. C.T.J. Dodson and K. Fekih. The effect of fibre orientation on paper formation. *J. Pulp Pap. Sci.* **17**(6):J203–J206, 1991

35. C.T.J. Dodson, A.G. Handley, Y. Oba and W.W. Sampson. The pore radius distribution in paper. Part I: The effect of formation and grammage. *Appita J.* **56**(4):275–280, 2003

36. C.T.J. Dodson, Y. Oba and W.W. Sampson. Bivariate normal thickness-density structure in real near-planar stochastic fibre networks. *J. Statist. Phys.* **102**(1/2):345–353, 2001

37. C.T.J. Dodson, Y. Oba and W.W. Sampson. On the distributions of mass, thickness and density in paper. *Appita J.* **54**(4):385–389, 2001

38. C.T.J. Dodson and W.W. Sampson. The effect of paper formation and grammage on its pore size distribution. *J. Pulp Pap. Sci.* **22**(5):J165–J169, 1996

39. C.T.J. Dodson and W.W. Sampson. Spatial statistics of stochastic fibre networks. *J. Statist. Phys.* **96**(1/2):447–458, 1999

40. C.T.J. Dodson and W.W. Sampson. Effect of correlated free fibre lengths on pore size distribution in fibrous mats. In **Advances in Paper Science and Technology**. *Trans. XIIIth Fund. Res. Symp.* (S.J. I'Anson, ed.), pp. 943–960, FRC, Manchester, 2005

41. C.T.J. Dodson and W.W. Sampson. Planar line processes for void and density statistics in thin stochastic fibre networks. *J. Statist. Phys.* **129**(2):311–322, 2007

42. C.T.J. Dodson and C. Schaffnit. Flocculation and orientation effects on paper formation statistics. *Tappi J.* **75**(1):167–171, 1992

43. M. Doi and S.F. Edwards. Dynamics of rod-like macromolecules in concentrated solution. Part 1. *J. Chem. Soc., Farad. Trans. 2* **74**(3):560–570, 1978

44. M. Doi and S.F. Edwards. Dynamics of rod-like macromolecules in concentrated solution. Part 2. *J. Chem. Soc., Farad. Trans. 2* **74**(5):918–932, 1978

45. N. Dooley and W.W. Sampson. Opportunities for improved light transmission formation analysis. *Pap. Tech.* **43**(8):31–36, 2002

46. C.B. Edwards and J. Gurland. A class of distributions applicable to accidents. *J. Am. Statist. Assoc.* **56**(295):503–517, 1961

47. S.Y. Eim, S.O. Hyuns, M.K. Kim, D.L. Lee and J.H. Park. New evaluation system for non-woven fabrics using image analysis technique: coverstocks, filters and interlinings. In proc. INDA-Tec. 96: 140–149, 1996

48. K.E. Evans and A.G. Gibson. Prediction of the maximum packing fraction achievable in randomly oriented short-fibre composites. *Compos. Sci. Tech.* **25**(2):149–162, 1986

49. R.R. Farnood, Sensing and modelling of forming and formation of paper. PhD Thesis, Department of Chemical Engineering and Applied Chemistry, University of Toronto, 1995

50. R.R. Farnood, C.T.J. Dodson and S.R. Loewen. Modelling flocculation. Part I: Random disk model. *J. Pulp Pap. Sci.* **21**(10):J348–J356, 1995

51. R.R. Farnood and C.T.J. Dodson. The similarity law of formation. In proc. *Tappi 1995 International Paper Physics Conference*, pp. 5–12, Niagara-on-the-Lake, Canada. Tappi Press, Atlanta, 1995

52. T.E. Farrington. Soft X-ray imaging can be used to assess sheet formation and quality. *Tappi J.* **71**(5):140–144, 1988

53. W. Feller. **An introduction to probability theory and its applications**. Third edition. John Wiley & Sons, New York, 1968

54. P.U. Foscolo, L.G. Gibilaro and S.P. Waldram. A unified model for particulate expansion of fluidised beds and flow in fixed porous media. *Chem. Eng. Sci.* **38**(8):1251–1260, 1983

55. F. Garwood. The variance of overlap of geometric figures with reference to a bombing problem. *Biometrika* **34**(1/2):1–17, 1947

56. E.I. George. Sampling random polygons. *J. Appl. Prop.* **24**(3):557–573, 1987

57. B. Ghosh. Random distances within a rectangle and between two rectangles. *Calcutta Math. Soc.* **43**(1):17–24, 1951

58. A. Goel, C.H. Arns, R. Holmstad, Ø.W. Gregersen, F. Bauget, H. Averdunk, R.M. Sok, A.P. Sheppard and M.A. Knackstedt. Analysis of the impact of papermaking variables on the structure and transport properties of paper samples by X-ray microtomography. *J. Pulp Paper Sci.* **32**(3):1–12, 2006

59. J. Görres and P. Luner. An apparent density model of paper. *J. Pulp Pap. Sci.* **18**(4):127–130, 1992

60. W.R. Goynes and K. Pusateri. Digital quantification of microscopic images to determine fiber orientation in nonwovens. *Microsc. Microanal.* **10**(2):1336–1337, 2004

61. K.-J. Gundström, P.O. Meinander, B. Norman, L. Reiner and T. Waris. High consistency former. *Tappi J.* **59**(3):58–61, 1976

62. L. Haglund, B. Norman and D. Wahren. Mass distribution in random sheets – theoretical evaluation and comparison with real sheets. *Svensk Papperstidn.* **77**(10):362–370, 1974

63. M. Hasuike, T. Kawasaki and K. Murakami. Evaluation of 3-D geometric structure of paper sheet. *J. Pulp Pap. Sci.* **18**(3):114–120, 1992

64. E.K. Hellén, M.J. Alava and K.J. Niskanen. Porous structure of thick fibe webs. *J. Appl. Phys.* **81**(9):6425–6431, 1997.

65. H. Higgins and J. de Yong. Visco-elasticity and consolidation of the fibre network during free water drainage. In **Consolidation of the Paper Web** *Trans. IIIrd Fund. Res. Symp.* (F. Bolam, ed.), pp. 242–268, BPBMA, London, 1966

66. R. Holmstad. Methods for paper structure characterisation by means of image analysis. PhD Thesis, Norwegian University of Science and Technology (NTNU), Trondheim, 2004

67. E. Holst and T. Schneider. Fibre size characterization and size analysis using general and bivariate lognormal distribution. *J. Aerosol Sci.* **16**(5):407–413, 1985

68. S. Huang, M. Goel, S. Ramaswamy, B.V. Ramarao and D. Choi. Transverse and in-plane pore structure characterisation of paper. *Appita J.* **55**(3):230–234, 2002.

69. T-Y. Hwang and C-Y. Hu. On a characterization of the gamma distribution: The independence of the sample mean and the sample coefficient of variation. *Annals Inst. Statist. Math.* **51**(4):749–753, 1999

70. A. Jena and K. Gupta. Pore structure characterization techniques. *Am. Ceram. Soc. Bull.* **84**(3):28–30, 2005

71. J.O. Johansson and O. Hössjer. A shot-noise model for paper fibres with non-uniform random orientations. *Scand. J. Statist.* **32**(3):351–363, 2005

72. P.R. Johnston, The most probable pore size distribution in fluid filter media. *J. Test. and Eval.* **11**(2):117–121, 1983

73. P.R. Johnston. Revisiting the most probable pore size distribution in filter media. The gamma distribution. *Filtrn. and Sepn.* **35**(3):287–292, 1998

74. O. Kallmes and H. Corte. The structure of paper, I. The statistical geometry of an ideal two dimensional fiber network. *Tappi J.* **43**(9):737–752, 1960. *Errata:* **44**(6):448, 1961

75. O. Kallmes, H. Corte and G. Bernier. The structure of paper, II. The statistical geometry of a multiplanar fiber network. *Tappi J.* **44**(7):519–528, 1961

76. O. Kallmes, H. Corte and G. Bernier. The structure of paper, V. The bonding states of fibres in randomly formed papers. *Tappi J.* **46**(8):493–502, 1963

77. O.J. Kallmes and G. Bernier. The structure of paper, VIII. Structure of idealized nonrandom networks. *Tappi J.* **47**(11):694–703, 1964

78. D.S. Keller and P. Luner. An instrument for electron beam and light transmission imaging of mass distribution in paper and fibrous webs. *Rev. Sci. Instr.* **69**(6):2495–2503, 1998

79. D.S. Keller and J.J. Pawlak. β-radiographic imaging of paper formation using storage phosphor screens. *J. Pulp Pap. Sci.* **27**(4):117–123, 2001

80. D.S. Keller and J.J. Pawlak. Analytical technique for the comparison of paper formation imaging methods. *J. Pulp Pap. Sci.* **27**(5):171–176, 2001

81. R.J. Kerekes, R.M. Soszynski and P.A. Tam Doo. The flocculation of pulp fibres. In **Papermaking Raw Materials**, *Trans. VIIIth Fund. Res. Symp.* (V. Punton, ed.), pp. 265–310, Mechanical Engineering Publications, London, 1985

82. R.J. Kerekes and C.J. Schell. Characterisation of fibre flocculation regimes by a crowding factor. *J. Pulp Pap. Sci.* **18**(1):J32–J38, 1992

83. R.J. Kerekes and C.J. Schell. Effects of fiber length and coarseness on pulp flocculation. *Tappi J.* **78**(2):133–139, 1995

84. R.J. Kerekes. Rheology of fibre suspensions in papermaking: An overview of recent research. *Nordic Pulp Pap. Res. J.* **21**(5):598–612, 2006

85. R.P. Kibblewhite. Effects of refined softwood:eucalypt pulp mixtures on paper properties. In **Products of Papermaking**, *Trans. Xth Fund. Res. Symp.* (C.F. Baker, ed.), pp. 127–167, Pira International, Leatherhead, 1993

86. A. Kiviranta and C.T.J. Dodson. Evaluating Fourdrinier formation performance. *J. Pulp Pap. Sci.* **21**(11):J379–J383, 1995

87. T. Komori and M. Itoh. A modified theory of fibre contact in general fiber assemblies. *Textile Res. J.* **64**(9):519–528, 1994

88. S. Koombhongse, L. Wenxia and D.H. Reneker. Flat polymer ribbons and other shapes by electrospinning. *J. Polym. Sci.* **39**(21):2598–2606, 2001

89. I. Kovalenko. A simplified proof of a conjecture of D.G. Kendall concerning shapes of random polygons. *J. Appl. Math. Stochastic Anal.* **12**(4):301–310, 1999

90. K.H. Lee, S. Givens, D.B. Chase and J.F. Rabolt. Electrostatic polymer processing of isotactic poly(4-methyl-1-pentene) fibrous membrane. *Polymer* **41**(23):8013–8018, 2006

91. D. Li and Y.N. Xia. Electrospinning of nanofibers: Reinventing the wheel? *Adv. Mater.* **16**(14):1151–1170, 2004

92. M. Li, Y. Guo, Y. Wei, A.G. MacDiarmid and P.I. Lelkes. Electrospinning polyaniline-contained gelatin nanofibers for tissue engineering applications. *Biomaterials* **27**(13):2705–2715, 2006

93. T. Li. Dependence of filtration properties on stainless steel medium structure. *Filtrn. and Sepn.* **34**(3):265–273, 1997

94. H. Linhart and M. Wilmot. Measuring the bivariate length-diameter distribution in samples of wool fibres. *Text. Res. J.* **34**:1107–1109, 1964

95. M. Lucisano and B. Norman. The forming and properties of quasi-random laboratory paper sheets. In proc. *Tappi 1999 International Paper Physics Conference*, pp. 331–340, San Diego. Tappi Press, Atlanta, 1999

96. Z.W. Ma, M. Kotaki, R. Inai and S. Ramakrishna. Potential of nanofiber matrix as tissue-engineering scaffolds. *Tissue Eng.* **11**(1-2):101–109, 2005

97. Z.W. Ma, M. Kotaki, T. Yong, W. He and S. Ramakrishna. Surface engineering of electrospun polyethylene terephthalate (PET) nanofibers towards development of a new material for blood vessel engineering. *Biomaterials* **26**(15):2527–2536, 2005

98. D.M. Martinez, K. Buckley, S. Jivan, A. Lindström, R. Thiruvengadaswamy, J.A. Olson, T.J. Ruth and R.J. Kerekes. Chracterizing the mobility of papermaking fibres during sedimentation. In **The Science of Papermaking**, *Trans. XIIth Fund. Res. Symp.* (ed. C.F. Baker), pp. 225–254, Pulp and Paper Fundamental Research Society, Bury, 2001

99. D.M. Martinez, H. Kiiskinen, A.-K. Ahlman and R.J. Kerekes. On the mobility of flowing papermaking suspensions and its relationship to formation. *J. Pulp Paper Sci.* **29**(10):341–347, 2003

100. S.G. Mason. Fibre motions and flocculation. *Tappi J.* **37**(11):494–501, 1954

101. J.A. Matthews, G.E. Wnek, D.G. Simpson and G.L. Bowlin. Electrospinning of Collagen Nanofibres. *Biomacromolecules* **3**(2):232–238, 2002

102. R.E. Miles. Random polygons determined by random lines in a plane. *Proc. Nat. Acad. Sci. USA* **52**:901-907,1157–1160, 1964

103. R.E. Miles. A heuristic proof of a long-standing conjecture of D.G. Kendall concerning the shapes of certain large random polygons. *Adv. in Appl. Probab.* **27**(2):397–417, 1995

104. L. Moroni, R. Licht, J. de Boer, J.R. de Wijn and C.A. van Blitterwijk. Fiber diameter and texture of electrospun PEOT/PBT scaffolds influence human mesenchymal stem cell proliferation and morphology, and the release of incorporated compounds. *Biomaterials* **27**(28):4911–4922, 2006

105. T. Nesbakk and T. Helle. The influence of the pulp fibre properties on supercalendered mechanical pulp handsheets. *J. Pulp Paper Sci.* **28**(12):406–409, 2002

106. W.K. Ng, W.W. Sampson and C.T.J. Dodson. The evolution of a pore size distribution in paper. In proc. Progress in Paper Physics Seminar: pp. 5–7, Stockholm, Sweden, 1996

107. K.J. Niskanen. Distribution of fibre orientations in paper. In **Fundamentals of Papermaking** (C.F. Baker and V.W. Punton, eds.) *Trans. IXth Fund. Res. Symp.*, pp. 275–303, Mechanical Engineering Publications, London, 1989

108. K. Niskanen. **Paper Physics**, Fapet Oy, Helsinki, 1998

109. K. Niskanen and H. Rajatora. Statistical geometry of paper cross sections. *J. Pulp Paper Sci.* **28**(7):228–233, 2002

110. B. Norman. Overview of the physics of forming. In **Fundamentals of Papermaking** (C.F. Baker and V.W. Punton, eds.) *Trans. IXth Fund. Res. Symp.*, Vol. III, pp. 73–149, Mechanical Engineering Publications, London, 1989

111. B. Norman, U. Sjödin, B. Alm, K. Björklund, F. Nilsson and J-L. Pfister. The effect of localised dewatering on paper formation. In proc. TAPPI 1995 International Paper Physics Conference, Niagara-on-the-Lake, 1995: pp. 55–59. Tappi Press, Atlanta, 1995

112. Y. Oba. Three dimensional structure of paper. PhD Thesis, Department of Paper Science, UMIST, 1999

113. A.G. Ogston. The spaces in a uniform random suspension of fibres. *Trans. Farad. Soc.* **54**:1754–1757, 1958

114. L. Onsager. The effects of shape on the interaction of colloidal particles. *Ann. N.Y. Acad. Sci.* **51**:627–659, 1949

115. D.H. Page, P.A. Tydeman and M. Hunt. A study of fibre-to-fibre bonding by direct observation. In **The Formation and Structure of Paper**, *Trans. IInd Fund. Res. Symp.* (F. Bolam, ed.), pp. 171–193, BPBMA, London, 1962

116. N. Pan. A modified analysis of the microstructural characteristics of general fibre assemblies. *Textile Res. J.* **63**(6):336–345, 1993

117. N. Pan. Fiber contact in fiber assemblies. *Textile Res. J.* **65**(10):616, 1995

118. N. Pan and W. Zhong. Fluid transport phenomena in fibrous materials. *Text. Prog.* **38**(2):1–93, 2006

119. A. Papoulis and S. Unnikrishna Pillai. **Probability, Random Variables and Stochastic Processes**. Fourth edition. McGraw Hill, Boston, 2002

120. J.G. Parkhouse and A. Kelly. The random packing of fibres in three dimensions. *Proc. Roy. Soc. Lond.* A **451**:737–746, 1995

121. Q.P. Pham, U. Sharma and A.G. Mikos. Electrospun poly(epsilon-caprolactone) microfiber and multilayer nanofiber/microfiber scaffolds: Characterization of scaffolds and measurement of cellular infiltration. *Biomacromolecules* **7**(10):2796–2805, 2006

122. A.P. Philipse. The random contact equation and its implications for (colloidal) rods in packings, suspensions, and anisotropic powders. *Langmuir* **12**(5):1127–1133, 1996. *Errata:* **12**(24):5971, 1996

123. A.P. Philipse and S.G.J.M. Kluijtmans. Sphere caging by a random fibre network. *Physica A* **274**(3-4):516–524, 1999

124. H.W. Piekaar and L.A. Clarenburg. Aerosol filters—Pore size distribution in fibrous filters. *Chem. Eng. Sci.* **22**(11):1399–1408, 1967

125. G.E. Pike and C.H. Seager. Percolation and conductivity: A computer study. I. *Phys. Rev. B.* **10**(4):1421–1434, 1974

126. B. Pourdeyhimi, R. Ramanathan and R. Dent. Measuring fiber orientation in nonwovens: Part I: Simulation. *Text. Res. J.* **66**(11):713–722, 1996

127. B. Pourdeyhimi, R. Ramanathan and R. Dent. Measuring fiber orientation in nonwovens: Part II: Direct tracking. *Text. Res. J.* **66**(12):747–753, 1996

128. B. Pourdeyhimi, R. Dent and H. Davis. Measuring fiber orientation in nonwovens: Part III: Fourier transform *Text. Res. J.* **67**(2):143–151, 1997

129. B. Pourdeyhimi and R. Dent. Measuring fiber orientation in nonwovens: Part IV: Flow field analysis *Text. Res. J.* **67**(3):181–187, 1997

130. B. Pourdeyhimi, R. Dent, A. Jerbi, S. Tanaka and A. Deshpande. Measuring fiber orientation in nonwovens: Part V: Real webs. *Text. Res. J.* **69**(3):185–192, 1999

131. B. Pourdeyhimi and H.S. Kim. Measuring fiber orientation in nonwovens: The Hough transform. *Text. Res. J.* **72**(2):803–809, 2002

132. B. Radvan, C.T.J. Dodson and C.G. Skold. Detection and cause of the layered structure of paper. In **Consolidation of the Paper Web** *Trans. IIIrd Fund. Res. Symp.* (F. Bolam, ed.), pp. 189–215. BPBMA, London, 1966

133. V. Ramakrishnan and D. Meeter. Negative binomial cross-tabulations, with applications to abundance data. *Biometrics* **49**(1):195–207, 1993

134. H.E. Robbins. On the measure of the random set. *Annal. Math. Statist.* **16**(4):342–347, 1945

135. S. Roberts and W.W. Sampson. The pore radius distribution in paper. Part II: The effect of laboratory beating. *Appita J.* **56**(4):281–283,289, 2003

272   References

136. S. Rolland du Roscoat. Contribution à la quantification 3D de réseaux fibreux par microtomographie au rayonnement synchrotron: applications aux papiers. PhD Thesis, EFPG/ESRF, Grenoble, 2007
137. S. Rolland du Roscoat, J.-F. Bloch and X. Thibault. Synchrotron Radiation microtomography applied to paper investigation. *J. Phys. D.* **38**(10A):A78–A84, 2005
138. S. Rolland du Roscoat, M. Decain, X. Thibault, C. Geindreau and J.-F. Bloch. Estimation of microstructural properties from synchrotron X-ray microtomography and determination of the REV in paper materials. *Acta Materialia* **55**(8):2841–2850, 2007
139. R. Salvado, J. Silvy and J.-Y. Dréan. Relationship between fibrous structure and spunbond process. *Text. Res. J.* **76**(11):805–812, 2006
140. W.W. Sampson. The structural characterisation of fibre networks in papermaking processes – A review. In **The Science of Papermaking**, *Trans. XIIth Fund. Res. Symp.* (ed. C.F. Baker), pp. 1205–1288, Pulp and Paper Fundamental Research Society, Bury, 2001
141. W.W. Sampson. Comments on the pore radius distribution in near-planar stochastic fibre networks. *J. Mater. Sci.* **36**(21):5131–5135, 2001
142. W.W. Sampson. A multiplanar model for the pore radius distribution in isotropic near-planar stochastic fibre networks. *J. Mater. Sci.* **38**(8):1617–1622, 2003
143. W.W. Sampson. A model for fibre contact in planar random fibre networks. *J. Mater. Sci.* **39**(8):2775–2781, 2004
144. W.W. Sampson and J. Sirviö. The statistics of inter-fibre contact in random fibre networks. *J. Pulp Pap. Sci.* **31**(3):127–131, 2005
145. W.W. Sampson, J. McAlpin, H.W. Kropholler and C.T.J. Dodson. Hydrodynamic smoothing in the sheet forming process. *J. Pulp Pap. Sci.* **21**(12):J422–J426, 1995
146. W.W. Sampson and S.J. Urquhart. The contribution of out-of-plane pore dimensions to the pore size distribution of paper and stochastic fibrous materials. *J. Porous Mater.* **15**(4):411–417, 2008
147. E.J. Samuelsen, Ø.W. Gregersen, P.J. Houen, T. Helle, C. Raven and A. Snigirev. Three dimensional imaging of paper by use of synchrotron X-ray microtomography. *J. Pulp Paper Sci.* **27**(2):50–53, 2001
148. C. Schaffnit. Statistical geometry of paper: modelling fibre orientation and flocculation. PhD Thesis, Department of Chemical Engineering and Applied Chemistry, University of Toronto, 1994
149. C. Schaffnit and C.T.J. Dodson. A new analysis of fibre orientation effects on paper formation. *Pap. ja Puu.* **76**(5):340–346, 1994
150. T. Schneider and E. Holst. Man-made mineral fibre size distributions utilizing unbiased and fibre length based counting methods and the bivariate lognormal distribution. *J. Aerosol Sci.* **14**(2):139–146, 1983
151. K. Schulgasser. Fiber orientation in machine made paper. *J. Mater. Sci.* **20**(3):859–866, 1985
152. O. Schultz-Eklund, C. Fellers and P.A. Johansson. Method for the local determination of the thickness and density of paper. *Nordic Pulp Paper Res. J.* **7**(3):133–139, 1992.
153. E. Schweers and F. Löffler. Realistic modelling of the behaviour of fibrous filters through consideration of filter structure. *Powder Tech.* **80**(3):191–206, 1994

154. V.I. Sikavitsas, G.N. Bancroft, J.J. Lemoine, M.A.K. Liebschner, M. Dauner and A.G. Mikos. Flow perfusion enhances the calcified matrix deposition of marrow stromal cells in biodegradable nonwoven fiber mesh scaffolds. *Ann. Biomed. Eng.* **33**(1):63–70, 2005

155. R.M. Soszyński. Simulation of two-dimensional nonrandom fibre networks. Oriented rectangles with randomly distributed centroids. *J. Pulp Pap. Sci.* **20**(4):J114–J118, 1994

156. R.M. Soszyński. Relative bonded area – A different approach. *Nord. Pulp Pap. Res. J.* **10**(2):150, 1995

157. Y.J. Sung, C.H. Ham, O. Kwon, H.L. Lee, D.S. Keller. Applications of thickness and apparent density mapping by laser profilometry. in: **Advances in Paper Science and Technology** (S.J. I'Anson, ed.), *Trans. XIIIth Fund. Res. Symp.*, pp. 961–1007, FRC, Manchester, 2005

158. J.C. Tan, J.A. Elliott and T.W. Clyne. Analysis of tomography images of bonded fibre networks to measure distributions of fibre segment length and fibre orientation. *Adv. Eng. Mater.* **8**(6):495–500, 2006

159. J.C. Tanner. The proportion of quadrilaterals formed by random lines in a plane. *J. Appl. Probab.* **20**(2):400–404, 1983

160. Y. Termonia. Permeability of sheets of nonwoven fibrous media. *Chem. Eng. Sci.* **53**(6):1203–1208, 1998

161. S. Toll. Packing mechanics of fiber reinforcements. *Polym. Eng. Sci.* **38**(8):1337–1350, 1998

162. M.M. Tomadakis and S.V. Sotirchos. Ordinary and transition regime diffusion in random fibre structures. *AIChE J.* **39**(3):397–412, 1993

163. H. Tomimasu, D. Kim, M. Suk and P. Luner. Comparison of four paper imaging techniques: β-radiography, electrography, light transmission, and soft X-radiography. *Tappi J.* **74**(7):165–176, 1991

164. C.M. van Wyk. Note on the compressibility of wool. *J. Textile Inst.* **37**:T285–T292, 1946

165. N.P. Vaughan and R.C. Brown. Observations of the microscopic structure of fibrous filters. *Filtrn. and Sepn.* **33**(8):741–748, 1996

166. X.Y. Wang and R.H. Gong. Thermally bonded nonwoven filters composed of bi-component polypropylene/polyester fiber. II. Relationships between fabric area density, air permeability and pore size distribution. *J. Appl. Polym. Sci.* **102**(3):2264–2275, 2006

167. J.F. Waterhouse. Effect of papermaking variables on formation. *Tappi J.* **76**(9):129–134, 1993

168. Y.B. Yi, L. Berhan and A.M. Sastry. Statistical geometry of random fibrous networks, revisited: Waviness, dimensionality and percolation. *J. Appl. Phys.* **96**(3):1318–1327, 2004

169. Z. Zhou, C. Chu and H. Yan. Backscattering of light in determining fibre orientation distribution and area density in nonwoven fabrics. *Text. Res. J.* **73**(2):131–138, 2003

170. L. Zhu, A. Perwuelz, M. Lewandowski and C. Campagne. Wetting behaviour of thermally bonded polyester nonwoven fabrics: the importance of porosity. *J. Appl. Polym. Sci.* **102**(1):387–394, 2006

Index